Getting Started with DuckDB

A practical guide for accelerating your data science, data analytics, and data engineering workflows

Simon Aubury

Ned Letcher

Getting Started with DuckDB

Group Product Managers: Kaustubh Mangrulkar and Apeksha Shetty

Publishing Product Manager: Deepesh Patel

Book Project Managers: Kirti Pisat and Hemangi Lotlikar

Senior Content Development Editor: Shreya Moharir

Technical Editor: Seemanjay Ameriya

Copy Editor: Safis Editing

Proofreader: Shreya Moharir

Indexer: Manju Arasan

Production Designer: Prashant Ghare

Senior DevRel Marketing Executive: Nivedita Singh

First published: June 2024

Production reference: 1240524

Published by Packt Publishing Ltd.

Grosvenor House

11 St Paul's Square

Birmingham

B3 1RB, UK.

ISBN 978-1-80324-100-5

www.packtpub.com

*To the many teachers who have inspired me, to my loving wife and children who have supported me,
and to Snowy the cat who has entertained me.*

– Simon Aubury

To Libby and Marcus. I am ever grateful for your love and support.

– Ned Letcher

Foreword

It's easy to convince someone that DuckDB is fast. Just run a query that aggregates a billion rows—you'll have some pretty compelling evidence about a second later. But showing someone how user-friendly it is? That takes some more time. Beyond all the effort that's gone into making DuckDB fast lies a second ocean of effort that's gone into making your life as a data wrangler easier—from the little helpers such as pattern matching on column names (so you can select all the columns that end in `'_price'`) to much bigger features such as being able to read directly from CSV files or JSON endpoints. DuckDB has so many affordances and hidden gems that you could spend years stumbling across them by chance, or by peering over the shoulder of a colleague. Luckily, you don't have to.

In this excellent book, Simon and Ned have combined the practicalities of what you need to know now with a wealth of hints and tips for getting the most out of DuckDB. Tips for doing more, much more easily. As I read through, I collected several pages of notes for things I'll start using immediately, and things to file away for future projects.

The chapter on DuckDB's extensions is particularly fruitful if you're looking to perform minor data miracles. You will learn how to pull raw data off of S3, chew through it in seconds, and export it into an Excel spreadsheet, instantly becoming the favorite data guru of an entire marketing department. (Although you may want to think twice before sealing that pact…)

– Kris Jenkins

Host of Developer Voices and Co-Founder of BullionVault

Contributors

About the authors

Simon Aubury has been working in the IT industry since 2000 as a data engineering specialist. He has an extensive background in building large, flexible, highly available distributed data systems. Simon has delivered critical data systems for finance, transport, healthcare, insurance, and telecommunications clients in Australia, Europe, and Asia Pacific. In 2019, Simon joined Thoughtworks as a principal data engineer and today is associate director of data platforms at Simple Machines in Sydney, Australia. Simon is active in the data community, a regular conference speaker, and the organizer of local and international meetups and data engineering conferences.

I want to thank the vibrant DuckDB community, which like many open source projects is built with the hard work, dedication, and passion of hundreds of contributors working together to build incredible things for us all.

Ned Letcher has worked as a data science and software engineering consultant since completing his PhD in computational linguistics in 2018 and currently works at Thoughtworks. He has designed and developed data-powered products and services across a range of industries and helped organizations and teams improve the effectiveness of their data processes and workflows. Ned has also worked as a Python trainer, supporting both tertiary students and data professionals across various organizations. He is active in the data community, speaking at and helping organize meetups and conferences, as well as contributing to a range of open source projects.

Thank you to everyone who's supported me on this odyssey, Dan and Lilly in particular: I couldn't have done this without you. Thanks to the technical advice and support from Craig Savage, Nathan Dines, and Elliana May. A big shoutout to the lovely folks at Code Black Brunswick, Ramblin' Man, and the Sporting Club Hotel, for keeping me fueled while I worked on my laptop in a fugue state.

About the reviewers

Torsten Grust is a professor of computer science at Universität Tübingen, Germany, where he has led the database systems research group since 2008. Torsten performs research into the design, compilation, optimization, and evaluation of a variety of database languages, modern dialects of SQL in particular. In this work, he often walks the fine line between database query and programming language technology. His group develops techniques that turn relational database systems into scalable processors also for non-relational query and programming languages. Torsten is at his happiest whenever he finds new evidence that database and programming language research can mutually benefit each other.

Louisa Lambrecht, **Björn Bamberg**, **Tim Fischer**, and **Denis Hirn** are research assistants in Torsten's group, where they tend to spend the lion's share of their (working) days dabbling with the intricacies of SQL processors and the innards of various database systems, including DuckDB.

Table of Contents

5

DuckDB Extensions   107

6

Semi-Structured Data Manipulation   125

12

DuckDB – The Wider Pond   311

Preface

There is no shortage of data being produced by humanity, in myriad formats, shapes, and ever-growing quantities. As it grows, so do the opportunities for leveraging data to benefit our world: improving decision making for governments, companies, and public organizations; supporting scientific research and technological advancements; and enabling the development of consumer products and important public services. To realize these opportunities, we are faced with an imperative: if we want to perform effective data analysis and develop products and services infused with machine learning, we must be able to manage, understand, and effectively work with the data that makes it possible.

Whether you are a data analyst, data scientist, research scientist, data engineer, software engineer, or data hobbyist, you are likely to face many of the same challenges when it comes to working with data. Analytical data workflows and applications require that data be loaded, cleaned, transformed, organized, exported, and crunched into summarized forms. A running joke amongst data practitioners is that they spend more time preparing and wrangling their data, as well as fighting with the tools that support their work than they do on the value-producing activities that are likely to be in their job descriptions. As data grows in volume and variety, these activities become both more difficult and more pressing to solve.

DuckDB is an analytical database that handles many of these challenges with ease. It enables data practitioners to streamline and improve the effectiveness of activities across the entire life cycle of data analysis and the development of analytical data infrastructure. It is simple to install and use on virtually any machine, running entirely in-process—without the overheads of connecting to and maintaining a dedicated server. At the same time, it offers blazing-fast performance for analytical operations, as well as powerful data management capabilities — features that are normally associated with distributed data processing engines and dedicated SQL database management systems. DuckDB's rich feature set makes it an incredibly versatile tool, being well suited to a range of different use cases, such as performing interactive data analysis and ad hoc data wrangling, efficiently querying data lakes, developing lean pipelines for transforming data, functioning as an operational data warehouse, and forming a low-latency query engine for powering responsive data apps. This versatility can also be a bit overwhelming at first, as it's hard to compare DuckDB with any one existing tool that you might be familiar with.

In this book, we'll dive into many of DuckDB's powerful and flexible capabilities. We'll give you a clear framework for how to think about what kind of a data tool DuckDB is and the types of applications it excels at. Through a range of hands-on examples, you'll learn how to make the most of this exciting tool and discover the many ways that you can incorporate it into your own analytical workflows and projects.

Who this book is for

If you're a data practitioner or student interested in an accelerated guide to understanding how to get the most out of DuckDB, this book is for you. This book is especially well suited for the data analyst who wants to explore complex and messy data, the data engineer who wants a lean and efficient transformation tool, or the data scientist who needs the flexibility of a data manipulation and management library that integrates seamlessly with Python and R.

DuckDB's primary interface is SQL, however, it is not the only way to interact with DuckDB databases, as you'll see in the chapters on Python and R. You'll benefit from having had prior experience working with SQL, however, we do include a brief SQL primer in *Chapter 1*, in case it's been a while between queries.

What this book covers

Chapter 1, An Introduction to DuckDB, starts our journey by introducing and positioning DuckDB in the data ecosystem. We unpack what type of database DuckDB is and identify use cases to which it is well suited and for which data practitioners are adopting it. We also set up the DuckDB client for use in our hands-on exploration, as well as spending some time on a brief SQL primer for those who could benefit from a refresher.

Chapter 2, Loading Data into DuckDB, provides an exploration of DuckDB's features, looking at how to load data into DuckDB from a range of formats and shapes, across CSV, JSON, and Parquet files.

Chapter 3, Data Manipulation with DuckDB, explores DuckDB's powerful data wrangling capabilities, focusing in particular on how we can use SQL operations and language features to perform common transformations used in data analysis.

Chapter 4, DuckDB Operations and Performance, dives into DuckDB operations and performance. We also learn how to leverage DuckDB performance optimizations for improving file-reading performance, through the use of Parquet files and Hive-partitioned data.

Chapter 5, DuckDB Extensions, introduces DuckDB's extension system, which enables users to extend DuckDB's functionality with useful features and capabilities that sit outside of the core DuckDB API. We cover how to install and load extensions, before getting hands-on with several extensions.

Chapter 6, Semi-Structured Data Manipulation, looks at a selection of DuckDB's data ingestion features for modeling, generating, manipulating, and querying semi-structured data. This also included covering DuckDB's strong support for working with JSON data.

Chapter 7, Setting up the DuckDB Python Client, sets up our Jupyter-Notebook-based IDE for working with DuckDB in Python and then unpacks the different ways to connect to DuckDB databases in Python.

Chapter 8, Exploring DuckDB's Python API, dives deeper into working with DuckDB's Python API, seeing how the DuckDB Python client is well suited for both analytical workflows and developing software packages that leverage DuckDB.

Chapter 9, Exploring DuckDB's R API, explores the features of DuckDB's R client, while exploring different ways you can leverage DuckDB in your R-based analytical workflows.

Chapter 10, Using DuckDB Effectively, surveys some of DuckDB's SQL enhancements and features that are designed to improve the experience of writing analytical queries.

Chapter 11, Hands-On Exploratory Data Analysis with DuckDB, dives deeper into the practical side of things, looking at how we can use DuckDB for performing exploratory data analysis with Python over a publicly available dataset.

Chapter 12, DuckDB – The Wider Pond, concludes our journey by exploring aspects of the wider DuckDB ecosystem, including tools and services you can use to supercharge workflows involving DuckDB—for both data analysis and software development—as well as online resources you can use to continue exploring and learning about DuckDB.

To get the most out of this book

We have tried to make the material in this book accessible to a wide range of readers, however, we will be assuming that readers have a foundational understanding of data concepts. Ideally, you will have experience working with SQL, however, we do provide a short SQL primer in *Chapter 1* for those in need of a refresher. We focus on using DuckDB with SQL for much of the book, though some chapters focus on using DuckDB with Python, and one using R. Having experience working with Python and R will help for these chapters, however, you should still find them accessible when diving into working with these languages for the first time.

Any common laptop or desktop machine should be suitable for working through the examples in this book.

Software/hardware covered in the book	Operating system requirements
The DuckDB CLI client	
The DuckDB Python client	Windows, macOS, or Linux
The DuckDB R client	

If you are using the digital version of this book, we advise you to type the code yourself or access the code from the book's GitHub repository (a link is available in the next section). Doing so will help you avoid any potential errors related to the copying and pasting of code.

Download the example code files

You can download the example code files for this book from GitHub at `https://github.com/ PacktPublishing/Getting-Started-with-DuckDB`. If there's an update to the code, it will be updated in the GitHub repository.

We also have other code bundles from our rich catalog of books and videos available at `https:// github.com/PacktPublishing/`. Check them out!

Conventions used

There are a number of text conventions used throughout this book.

`Code in text`: Indicates code words in text, database table names, data types, folder names, filenames, file extensions, pathnames, and user input. Here is an example: "To apply a filter to our SQL query, we can use a WHERE clause."

A block of code is set as follows:

```
CREATE TABLE foods (
    food_name VARCHAR PRIMARY KEY,
    color VARCHAR,
    calories INT,
    is_healthy BOOLEAN
);
```

When we wish to draw your attention to a particular part of a code block, the relevant lines or items are set in bold:

```
SELECT food_name, color
FROM foods
WHERE food_name = 'apple';
```

Bold: Indicates a new term, an important word, or words that you see onscreen. For instance, words in menus or dialog boxes appear in **bold**. Here is an example: "You'll need to select the **CSV** option from the **Export** button."

> Tips or important notes
> Appear like this.

Get in touch

Feedback from our readers is always welcome.

General feedback: If you have questions about any aspect of this book, email us at `customercare@packtpub.com` and mention the book title in the subject of your message.

Errata: Although we have taken every care to ensure the accuracy of our content, mistakes do happen. If you have found a mistake in this book, we would be grateful if you would report this to us. Please visit `www.packtpub.com/support/errata` and fill in the form.

Piracy: If you come across any illegal copies of our works in any form on the internet, we would be grateful if you would provide us with the location address or website name. Please contact us at `copyright@packtpub.com` with a link to the material.

If you are interested in becoming an author: If there is a topic that you have expertise in and you are interested in either writing or contributing to a book, please visit `authors.packtpub.com`.

Share Your Thoughts

Once you've read *Getting Started with DuckDB*, we'd love to hear your thoughts! Scan the QR code below to go straight to the Amazon review page for this book and share your feedback.

https://packt.link/r/1-803-24100-4

Your review is important to us and the tech community and will help us make sure we're delivering excellent quality content.

Download a free PDF copy of this book

Thanks for purchasing this book!

Do you like to read on the go but are unable to carry your print books everywhere?

Is your eBook purchase not compatible with the device of your choice?

Don't worry, now with every Packt book you get a DRM-free PDF version of that book at no cost.

Read anywhere, any place, on any device. Search, copy, and paste code from your favorite technical books directly into your application.

The perks don't stop there, you can get exclusive access to discounts, newsletters, and great free content in your inbox daily

Follow these simple steps to get the benefits:

1. Scan the QR code or visit the link below

https://packt.link/free-ebook/978-1-80324-100-5

2. Submit your proof of purchase
3. That's it! We'll send your free PDF and other benefits to your email directly

1

An Introduction to DuckDB

Data is everywhere, stored in a huge variety of systems across many different formats, and with an ever-growing number of tools available to data practitioners to practice their craft. **DuckDB** is a relatively new and explosively popular **database management system (DBMS)** that is increasingly being adopted for analytical data workloads by data scientists, data analysts, data engineers, and software engineers. DuckDB is open source software that is made available under the permissive MIT license, making it friendly to both commercial and non-commercial applications alike. The non-profit *DuckDB Foundation* stewards the long-term health of the DuckDB project, and the development of DuckDB is supported by DuckDB Labs, which employs the project's core contributors.

In this chapter, we'll unpack what type of database DuckDB is and identify use cases that DuckDB is well suited to and that data practitioners are increasingly adopting it for. We'll also outline the different deployment options DuckDB comes with and take you through how to install it on your own system so that you're ready to dive into the hands-on examples in this book. Finally, we'll go through a quick primer on **Structured Query Language (SQL)**, the query language DuckDB uses for its primary interface that we'll be using for many of the exercises in this book. If you've wrangled your fair share of SQL before, you may want to just skim through this section. If you're newer to using SQL, or it's been a while between queries, then you'll want to dive into these hands-on exercises.

By the end of this chapter, you'll be able to orient DuckDB within the landscape of data tooling and understand what kinds of use cases you may want to consider leveraging it for, as well as be able to recognize when other data processing tooling may be more appropriate.

Across the rest of the book, we'll show you how to take DuckDB through its paces, and in doing so, hopefully impart a sense of why there is so much enthusiasm around it. Right now, let's jump into setting the scene for our DuckDB explorations by covering the following topics:

- What is DuckDB?
- Why use DuckDB?
- DuckDB deployment options and installation
- A short SQL primer

Technical requirements

To follow along with the examples in this book, you'll need access to a computer running either Windows, macOS, or Linux, and an internet connection to download and then install DuckDB. In later chapters, you'll also need to download some datasets that we'll be using to explore DuckDB's analytical capabilities. The examples we present are available for you to access in this GitHub repository: `https://github.com/PacktPublishing/Getting-Started-with-DuckDB`.

What is DuckDB?

Whether you're an experienced data practitioner or just getting started working with data, you will almost certainly find yourself having to navigate the dizzying number of databases and data processing tools that you can choose from to support data-centric applications and operational systems. The reason for this overwhelming choice is that when it comes to data processing and management architectures, there is no one-size-fits-all. Each tool necessarily comes with its own set of trade-offs that make it well suited to a particular flavor of application and less so to others.

With that in mind, let's dig into what kind of database DuckDB is and where it sits in the data-tooling landscape so that we can unpack what kinds of applications and use cases it is well suited to. One description of DuckDB, which you might encounter when poking around online resources, is the following:

DuckDB is an in-process SQL OLAP DBMS.

While this is a fairly dense description, invoking several distinct concepts from the world of databases and software applications, it does a great job of positioning where DuckDB sits in relation to other databases and data processing tools. So, let's break this description down, going through each component and working our way from right to left:

- A **database management system** (**DBMS**) is a software application for managing structured data in a database, allowing users and applications to store, manipulate, delete, and query records. While you might hear the term *database* being used as shorthand for *DBMS*, it's worth noting that a DBMS provides additional functionality on top of the core features of a database—which is essentially to store data in a structured format that supports efficient retrieval and manipulation. A DBMS provides an interface between the database and its users, enabling them to effectively create, read, update, and delete data, while also managing the integrity, scalability, and security of the database. DuckDB is a fully-fledged DBMS that manages all these concerns for users.

- **Online analytical processing** (**OLAP**) is a data processing paradigm that is characterized by complex queries over large volumes of multidimensional data, which often involve processing significant portions of a dataset. These analytical workloads often involve applying column-wise aggregation functions over entire tables and joining large tables together. The term was created in contrast to **online transaction processing** (**OLTP**), which describes transaction-oriented DBMS tools, such as PostgreSQL, MySQL, and SQLite, which are typically used as operational databases supporting software applications, where frequent reading and writing of individual records is the dominant access pattern. DuckDB is designed and optimized for fast and efficient performance over OLAP workloads.

- **SQL** is a popular programming language used for storing, manipulating, and querying records in a wide variety of databases and data stores. It is a standard interface used for interacting with and managing relational databases, which are databases characterized by the representation of data as tables of rows and columns, with formal relationships defined across tables. SQL's increasing ubiquity has made it something of a de facto choice for code-defined data-querying interfaces. DuckDB has its own SQL dialect, which forms the primary interface for interacting with DuckDB databases. As we will see, there are also non-SQL interfaces available for users to work with DuckDB databases. In the last section of this chapter, *A short SQL primer*, we'll cover a brief introduction to the fundamentals of working with SQL for those who are new to working with it or a little rusty.

- **In-process** means that DuckDB runs embedded within a host process. This is in contrast to most DBMSs, which typically operate standalone, running in a separate process from consuming applications, often on a remote server. By adopting an in-process model rather than a client-server architecture, DuckDB greatly simplifies installation and integration, removing the need to install and manage a standalone DBMS service, as well as the need to connect and authenticate with a remote server. A notable example of an in-process DBMS that you may have encountered is SQLite, which is a popular choice for software developers distributing apps that require reading and writing local transactional data, such as user data for mobile apps and lightweight web apps.

Putting all these pieces together, we can see that DuckDB is a fully featured **relational DBMS (RDBMS)** that is designed for analytical workloads, provides a SQL interface, and runs entirely embedded in a host process.

When compared with other popular databases, DuckDB is perhaps most similar to the ubiquitous SQLite in that they are both simple in-process DBMSs that write to a single-file storage format, and they are also both free and open source. The key difference between the two tools is that SQLite is optimized for row-oriented OLTP workloads and hence does not perform well on complex analytical workloads, whereas DuckDB is purpose-built for these workloads, offering extremely good performance over them. It's for this reason that DuckDB is sometimes described as SQLite for OLAP. In fact, DuckDB appears to be the first production-ready in-process OLAP DBMS.

In the next section, we'll explore the reasons why people are increasingly adopting DuckDB and finding it to be a valuable workhorse in their analytical data toolkit.

Why use DuckDB?

So, why might you want to use DuckDB? Let's start by zooming all the way out. As a data practitioner, there are two broad contexts where you might find yourself getting excited about leveraging DuckDB:

1. Using DuckDB to scale and supercharge data science, data analytics, and ad hoc data-wrangling workflows.

2. Using DuckDB as a building block to build operational data engineering infrastructure and interactive data products.

The first of these is likely to be of interest to data practitioners with analytical workflows, such as data scientists, data analysts, and machine learning engineers, whereas the second is more likely to be relevant to data engineers and machine learning engineers building data infrastructure, as well as software engineers building user-facing data products. In this book, we'll be focusing more on using DuckDB to supercharge analytical workflows; however, if you're looking to use DuckDB for building operational data infrastructure and data products, this book will still be a great starting point to get you up to speed with DuckDB's capabilities that make it well suited to these kinds of applications.

In this section, we'll first go through some use cases that land in DuckDB's sweet spot, before looking at DuckDB's features that make it especially well-suited to these applications. We'll finish up by discussing contexts where other tools may be more appropriate.

DuckDB use cases

DuckDB is an incredibly versatile tool for analytical data processing and management, so any attempt to describe its full range of potential applications will almost certainly be incomplete. To give you a sense of the flavor of possible applications, we'll go through a range of use cases for DuckDB across the two broad categories mentioned previously: analytical workflows and building operational data infrastructure and products.

Supporting analytical workflows

A major component of the workflows of data scientists and data analysts is activities that involve processing often quite large datasets, from cleaning data, transforming data into the right shape, structured data modeling, running statistical algorithms, and training machine learning models. If you talk to a practitioner who has been in the trenches for a while, they will likely tell you that sometimes they feel like they're spending more time fighting the tools they use for these tasks than they are being productive. Often the size of the data is a limiting factor, with many popular data processing tools, such as pandas dataframes in Python and dataframes in R, simply not being able to handle the size of target datasets within the memory of your workstation or, if they can, taking a frustrating amount of time to process.

Once you've hit the limits of your local machine, conventional wisdom is that you need to take your workload to a distributed data compute framework such as Apache Spark, Dask, or Ray, or perhaps ingest your data into a cloud data warehouse or a data lake, where a distributed SQL query engine such as Google BigQuery, Trino, or Amazon Athena can be used to run queries at scale. These solutions significantly increase the complexity of your workflows, requiring complex supporting infrastructure that must be managed and maintained, with a hefty price tag often associated with such managed services. If you're lucky enough to have access to these tools, they still come with additional challenges, such as working with unfamiliar or constrained interfaces, and when things go wrong, you may often find yourself having to debug arcane and confusing stack traces from the underlying compute engine.

This is where DuckDB can come to the rescue, offering the simplicity of an in-process tool, with a familiar SQL interface (as well as non-SQL interfaces if you prefer) that is optimized for running complex OLAP queries over large datasets. Not only is DuckDB blazingly fast, but it is also able to handle out-of-core workloads (datasets that don't fit into memory), enabling you to scale your workflows on a single machine much further before you need to consider more complex distributed data processing solutions.

In recent times, there have been developments in dataframe libraries that help address the performance limitations of tools such as pandas dataframes and R dataframes, such as Dask and Modin, which allow you to perform simple parallelization of dataframe operations across your CPU cores, as well as providing on-ramps to run the same queries across a distributed cluster. We also have dataframe libraries such as Polars and Vaex, which are built on top of Apache Arrow, providing more efficient memory utilization, parallelization, and the ability to handle some out-of-core workloads. These innovations in the data ecosystem are pleasing to see; however, these tools are still ultimately dataframe tools, focusing primarily on querying and data transformation—they do not give you the data management features of a DBMS.

By virtue of being a fully-fledged DBMS, DuckDB provides a range of affordances that data practitioners may not realize they're missing from their current analytical processing workflows:

- DuckDB provides transactional guarantees through **ACID** properties (**atomicity**, **consistency**, **isolation**, and **durability**), meaning that you don't have to worry about corrupted data if your Python or R process crashes midway through a job.

- Data integrity guarantees can be enabled through the specification of constraints that enforce properties over data inserted into tables. DuckDB allows you to specify PRIMARY KEY constraints, which enforce uniqueness across rows within a table, FOREIGN KEY constraints, which enforce referential integrity for relationships across tables, and NOT NULL constraints over column values. DuckDB also provides the ability to apply arbitrary CHECK constraints to column values in the form of Boolean expressions, such as ensuring that string identifiers only contain alphanumeric characters.

- While you can use DuckDB as an entirely in-memory database, its database can also be persisted to disk and used across processes, even allowing multiple processes to read concurrently. This enables workflows and consuming patterns that dataframe libraries cannot readily support on their own.

- DuckDB also includes a rich suite of data integrations, with an eye toward performance. Notable examples include optimized CSV, Parquet, and JSON loaders, which can read files in parallel, the ability to read Hive-partitioned files, and the PostgreSQL, MySQL, and SQLite extensions, which allow DuckDB to query directly from source tables in external databases, rather than having to rely on bulk imports that must be periodically refreshed.

When a data team starts to hit the limits of their existing tooling, whether due to missing data management features or insufficient performance, it's common for the team to start building out their own bespoke tools and packages. Since this kind of custom tooling is typically not core to the value the team is providing, these resources can suffer from defects due to insufficient resources being able to be dedicated to their development. Using a well-maintained and tested DBMS that is optimized for analytical workloads removes the busywork that is associated with maintaining tooling that doesn't represent your core value proposition.

DuckDB's powerful feature set makes it a versatile tool for a range of analytical workflows, whether you're performing **exploratory data analysis** (**EDA**), quickly transforming between common data formats, or building complex data science pipelines. DuckDB enables you to slurp up large datasets from across heterogeneous data sources, with a rich set of features for cleaning dirty data and normalizing inconsistent schema, through a simple interface and with blazing performance. DuckDB also has great integrations with familiar analytical tools commonly used in the data ecosystem, allowing you to mix and match DuckDB with complementary tools to assemble your own effective workflows. For example, DuckDB can query directly from and write to pandas and Polars dataframes and R dataframes, as well as Apache Arrow tables. It also offers the ability to use alternative query interfaces to SQL that may be more familiar to data scientists and data analysts, such as `dplyr` in R and Ibis in Python. In addition to being a powerful workhorse for complex analytical queries, all this versatility makes DuckDB a valuable Swiss Army knife that is worth having in your analytical data toolkit.

Finally, data scientists and data analysts often find themselves building custom interactive data apps or dashboards for use as **proof of concepts** (**POCs**), bespoke tools that support common workflows, or for publishing internal decision-support tools within their organization. Powerful open source dashboarding tools such as R Shiny, Streamlit, and Plotly Dash streamline the development of such data apps; however, they typically leave the integration of a data source up to the developer. DuckDB is a perfect complement to these tools, offering simple in-process integration with no external dependencies and enabling fast analytical querying performance, which is important for low-latency response times that improve the user experience of your data apps. We unpack this particular application of DuckDB further in the *DuckDB-powered data apps* section in *Chapter 12*.

Building data infrastructure

While much of the explosive growth and excitement that DuckDB has seen has been driven by folks adopting it for the types of analytical workflows we have just discussed, there is another area of application that is starting to show increased amounts of activity and demand for, which draws upon similar themes of doing more with less and simplifying and streamlining workloads. This sees DuckDB being used as a building block in modern data infrastructure for use cases that involve small-to-medium data rather than truly big data, as well as use cases that require low-latency responsiveness for consumer-facing interactive data apps. Common to these applications is a shift away from the paradigm of moving your compute to the data, which is seen as the conventional wisdom for effectively working with big data, and a move toward bringing your data to the compute. For smaller data workloads, this can be faster, more efficient, and cheaper to build and maintain.

Much of the development in modern data processing technologies has been dominated by the needs of hyperscale organizations, with scale-out tools such as MapReduce, Hadoop, and Apache Spark, as well as cloud data warehouses such as Snowflake and BigQuery, dominating the landscape. Most organizations, however, do not operate at hyperscale, and oftentimes data processing needs are quite moderate in comparison. The cleaned and enriched datasets that drive the modern data-informed business—providing **business intelligence** (**BI**) across sales, marketing, growth, and product innovation—tend not to reach the petabyte scale. There is an opportunity for data teams in many organizations to adopt leaner data architectures that are optimized for more moderate data workloads and that come with the benefits of reduced complexity and much lower total cost of ownership.

DuckDB's performance characteristics make it well placed to be a core building block in such architectures. Some examples include the following:

- Using DuckDB to perform transformations in **extract, transform, and load** (**ETL**) pipelines as an alternative to tools such as Apache Spark. Compute instances can be spun up on demand, invoking DuckDB to pull down and transform data.

- For data lake contexts, where structured and semi-structured data has been landed in object storage, DuckDB can be used as a lightweight alternative to distributed SQL query engines, which data teams might otherwise reach for, such as Google BigQuery, Amazon Athena, Trino, and Dremio.

- For some scenarios, DuckDB also offers the potential to replace the use of cloud data warehouses such as Snowflake or OLAP engines such as ClickHouse, where utilization of these powerful resources would be low. If your organization is only consuming a handful of data sources to produce conformed tables that drive a small number of reporting use cases, then using DuckDB to build small, targeted data cubes may well be sufficient for your needs.

Some folks have already started to roll their own solutions for adopting these architectures. See, for example, the *Modern Data Stack in a Box with DuckDB* post by Jacob Matson (`https://duckdb.org/2022/10/12/modern-data-stack-in-a-box.html`), which explores the use of open source tools to create an end-to-end lightweight data stack, with DuckDB at its core. Another post, *Build a poor man's data lake from scratch with DuckDB*, by Pete Hunt and Sandy Ryza (`https://dagster.io/blog/duckdb-data-lake`), explores using DuckDB as a SQL query engine on top of a data lake. Meanwhile, there are also companies emerging that are oriented around offering hosted platforms that provide serverless analytics platforms driven by DuckDB, the most notable example being MotherDuck (`https://motherduck.com`).

Another area where traditional scale-out approaches to data processing have shown to be not always fit for purpose is around interactive data applications, such as BI dashboards and bespoke data apps. In such applications, low-latency query results in response to user interaction are crucial for supporting dynamic and ad hoc workloads with a positive user experience. However, most cloud data warehouses and distributed data processing engines are simply not able to provide the low-latency response times required for these types of workloads and must be augmented with different types of pre-aggregation and caching strategies, often in the form of separate service, which further increases complexity and architectural surface area. DuckDB's blazing fast speeds over analytical workloads make it a compelling choice for being the backing query engine for interactive data applications. For example, the hosted BI service Mode recently switched to using DuckDB as their in-memory query engine in order to improve the speed of queries (`https://mode.com/blog/how-we-switched-in-memory-data-engine-to-duck-db-to-boost-visual-data-exploration-speed`). Hex and Observable are two hosted data analytics notebook services offering rich visualizations and interactivity that both recently added DuckDB integration to supercharge users' workflows. Another notable example is Rill Data Developer, an open source tool for building dashboards, which is built on DuckDB to provide rapid response times for queries.

The use of DuckDB as a building block for data infrastructure and interactive data applications is a notable emerging trend and one we think is worth paying attention to. In the next section, we'll further unpack the features of DuckDB that serve to make it appealing for both analytical workflows and building operational data infrastructure and data products.

DuckDB features

You may find yourself asking, what makes DuckDB so well suited to scaling analytical workflows and being used as a building block in data infrastructure? Here are some key features of DuckDB that have led to it increasingly being adopted by data practitioners.

Performance

DuckDB is optimized for OLAP workloads, making it blazingly fast and efficient for the kinds of queries frequently seen in analytical workflows. It achieves this through a range of design choices and optimizations:

- As with most modern OLAP engines, DuckDB employs a column-based execution model to enable better performance over operations that are characteristic of analytical workloads. DuckDB uses a highly tuned vectorized query engine that works by processing chunks of columns at a time. Operating on column chunks rather than entire columns means that queries involving multiple operations, which require intermediate aggregations, are less likely to result in **out-of-memory** errors. The chunks are also tuned to be small enough so that they remain inside the CPU's ultra-low latency L1 cache—the CPU's fasted dedicated memory, which is drastically faster than main memory.

- DuckDB leverages a range of compression algorithms, which exploit similarities in values within columns, to reduce its storage size on disk, which in turn improves read times.

- DuckDB employs an end-to-end query optimizer. This means that rather than executing queries as they are written, DuckDB can automatically rewrite queries to be much more efficient as they are written, DuckDB can automatically rewrite queries to be much more efficient.

- Almost all of DuckDB's operations come with automatic parallelism, allowing it to distribute operations over multiple CPU threads, resulting in reduced processing time.

DuckDB is also able to support **out-of-core** workloads, where the data to be processed does not fit within available memory. It does this by spilling over into temporary disk storage when memory is exhausted. This does increase processing times due to the slower read times of persistent storage compared to memory; however, this is typically preferable to the query failing outright. These costs can also be mitigated by the selective use of low-latency SSD drives for applications where this is a concern.

Ease of use

The design choice of operating in-process means that users of DuckDB don't need to concern themselves with installing, maintaining, and authenticating with a standalone database server. Another key design decision of DuckDB was for it not to make use of any third-party dependencies. This makes DuckDB extremely portable across platforms and has also enabled DuckDB to be made available for a wide range of languages and runtimes. This feature of DuckDB has increased its accessibility to a diverse range of consumers, allowing it to be readily incorporated into a wide variety of workflows and tech stacks.

DuckDB also has a strong focus on improving the ergonomics of working with SQL. It has a PostgreSQL-like SQL dialect, making it familiar to many data practitioners, and also includes a wide range of alternative function aliases, matching names used in other popular databases that many practitioners will be familiar with. Notably, DuckDB's SQL dialect has a range of enhancements designed to improve productivity when writing analytical SQL queries. Some of these include the following:

- Automatic casting of data types where possible, which serves to simplify SQL queries.

- Simple creation of `LIST` and `STRUCT` data types using literal values.

- Accessing attributes of `STRUCT` data types using dot notation.

- Simple string and list slicing syntax similar to Python.

- The ability to define anonymous lambda functions within SQL queries that can be used for transforming and filtering lists.

- List comprehension syntax similar to Python's, enabling more convenient list transformation and filtering.

- Improved ergonomics for column selection using `SELECT *`, with the `EXCLUDE` and `REPLACE` clauses, which allow you to leverage wildcard selection for more concise queries, while still being able to exclude or transform specific columns. This pattern-matching power can be further enhanced with the COLUMNS () expression, which allows you to select columns that match a regular expression or the output of a lambda function.

- Column aliases that can be reused by subsequent columns in the same SELECT statement, as well as in WHERE and ORDER BY clauses.

- The ability to start SELECT statements with its FROM clause, allowing you to improve the readability of queries by frontloading the data source. Additionally, omitting the SELECT clause entirely is interpreted as having specified SELECT *, making this common query pattern more concise.

- Function-call chaining within SQL queries, similar to familiar data processing APIs such as that of pandas, which is designed around method-call chaining.

- Trailing commas in SQL queries.

We'll cover some of these features in *Chapter 10*. For a more comprehensive treatment of the **friendly SQL** enhancements that DuckDB's SQL dialect provides, see the DuckDB documentation: https://duckdb.org/docs/guides/sql_features/friendly_sql.

Versatility

DuckDB comes enabled with a range of affordances that make it an incredibly versatile data processing and management tool. In terms of data sources, DuckDB can read and write data formats that are the mainstays of the data world: Parquet, CSV, and JSON (including newline-delimited JSON). In addition to reading from disk, these files can be read from remotely hosted files, and DuckDB can both read and write cloud object storage services using the **Simple Storage Service (S3)** API. DuckDB is also able to directly connect to and query from multiple databases at a time, including PostgreSQL, MySQL, and SQLite databases.

DuckDB also has tight integrations with in-memory data formats that are commonly used in the analytical data ecosystem, including pandas and Polars dataframes and R dataframes, as well as Apache Arrow tables. DuckDB provides the ability to query directly from these in-memory data structures, as well as export to them.

DuckDB's rich set of official clients also means that it can be used in a range of contexts beyond Python- and R-based workflows. Of particular note here is the DuckDB **WebAssembly (Wasm)** client, which enables developers to readily integrate DuckDB into web apps that can be published to the internet for anyone to access with a browser.

Together, all these capabilities make DuckDB a veritable data Swiss Army knife for working with analytical data, opening up many interesting applications that we have only just scratched the surface of.

Powerful analytics capabilities

DuckDB offers extensive support for composing complex queries through SQL, with a particular emphasis on features commonly used in analytical workloads. DuckDB has support for optimized aggregation and join operations, column indexes, window functions, and full-text search. DuckDB supports a wide range of functions for working with different types of data, including numeric operations, date and time operations, date formatting, text operations, bit strings, and string pattern matching, including regular expressions. Analytical workloads are further enabled by SQL commands such as `SAMPLE`, which provides convenient probabilistic sampling of datasets, the `PIVOT` command for creating pivot tables, `ASOF` joins for interpolating missing values when joining time series data, and the `QUALIFY` clause, for convenient filtering of window-functions results.

DuckDB also has a rich array of data types, which, in addition to those you'd expect, such as numeric, date, and text types, also includes handy types such as `INTERVAL` for date ranges, `ENUM` for enumerations, and powerful nested data types for holding multiple values, which include `ARRAY`, `LIST`, `STRUCT`, `MAP`, and `UNION`. DuckDB also offers support for analytical applications that involve working with geospatial data, using its spatial extension.

As this book was nearing completion, DuckDB released a vector similarity search extension, which enables using DuckDB's `ARRAY` data type for machine learning and data science applications involving **vector embeddings** and **semantic similarity search**.

Engaged community

DuckDB is open sourced under the permissive MIT license, making it readily adoptable and extensible for a wide range of commercial and non-commercial applications. The community that has formed around DuckDB has generated many valuable open source extensions, integrations, and alternative DuckDB clients for different languages and runtimes. This rich ecosystem of DuckDB-oriented projects is indicative of and has been a powerful catalyst for the enthusiasm behind DuckDB.

When is DuckDB not the right tool?

As we have already discussed, when it comes to databases, there is no one-size-fits-all solution. This means that DuckDB won't always be the right tool for the job.

Since it is optimized for analytical workloads running on a single machine, DuckDB has been intentionally designed to perform well under a specific set of access patterns, which you should confirm are acceptable for your use case. If your workloads correspond more to the OLTP paradigm, with many individual record transactions, including frequent writes, then DuckDB, which is optimized for OLAP workloads consisting mostly of read access and column-oriented queries, is likely not the best tool for you. If you're after an in-process DBMS that's optimized for OLTP workloads, then SQLite is hard to beat. Another specific consideration is that in order to open a DuckDB database that can be written to, only one process can both read from and write to the database. If you need multiple processes to be able to concurrently read from a DuckDB database, you must load it in read-only mode.

DuckDB's performance is truly impressive, allowing you to push the limits of what a single machine can do, arguably redefining what actually constitutes big data for analytical workloads. However, it is ultimately a database that operates in-process on a single machine, and so there are limits to how far it can be pushed. If your target workloads involve regularly processing petabyte-scale datasets, then you will likely need to use some form of distributed data processing-enabled platform.

Now that we've established what DuckDB is and when it makes sense to use it, we'll start to get more practical by looking at DuckDB's deployment options and how you can get started working with it.

DuckDB deployment options and installation

In this section, we'll look at ways you can use and integrate DuckDB into your analytical workflows, operational infrastructure, and data products. We'll start by outlining the different clients available for DuckDB, and then we'll go through how to get started working with the DuckDB **command-line interface (CLI)** on your own machine.

DuckDB deployment options

The data technology landscape is a big place, full of a diverse range of data practitioners with different skill sets and a wide range of tools built on a heterogeneous set of technologies. To cater to this diversity, DuckDB is made available via a number of different official client APIs, covering users of Python, R, JavaScript, Rust, Swift, Julia, Java, C, and C++. In the case of JavaScript, there are two clients: one for Node.js, oriented around backend applications, and one for Wasm, which allows DuckDB to run entirely within any modern web browser. Additionally, DuckDB is also made available as a cross-platform CLI, which is distributed as an executable that can be run virtually anywhere you have access to a command line. At the time of writing, there are also community-supported clients for Rust, Go, C#, Ruby, Common Lisp, Zig, and Crystal. As this list is ever-growing, we encourage you to consult the DuckDB documentation for an up-to-date list of official DuckDB clients and third-party contributed clients: `https://duckdb.org/docs/api/overview`.

This large selection of clients and integrations provides good coverage over languages and runtimes that meet the needs of a wide range of data practitioners. Here are the clients we'll cover in this book:

- **The DuckDB CLI** is a cross-platform executable that gives you a CLI for interacting with DuckDB databases via SQL. It's compiled without any external dependencies, meaning that you can run it virtually anywhere you have access to a terminal. Its ease of installation and portability make it a great way to get started with DuckDB, which is why we will be making use of it for many of the hands-on examples in this book. We'll walk through how to install it and get started using it shortly.

- **The DuckDB Python client** enables Python processes to readily communicate with DuckDB databases. It offers a number of distinct APIs for querying and interacting with DuckDB, making it suitable for a range of applications, spanning the spectrum of interactive data analysis to building data infrastructure and analytical data products. In *Chapter 7*, we introduce the

DuckDB Python client, focusing in particular on how to connect to DuckDB databases. In *Chapter 8*, we go on a deeper dive into the DuckDB Python client, focusing on two of the APIs that it exposes: the DuckDB-specific Relational API and the Python **Database API (DB-API)**. In *Chapter 11*, we'll use the Python client again, this time to perform EDA of a dataset in a Jupyter Notebook. Finally, in *Chapter 12*'s *Alternative DuckDB query interfaces* section, we touch on DuckDB's Spark API—another component of the DuckDB Python client—which enables interacting with DuckDB databases using PySpark queries.

- **The DuckDB R client** enables R sessions to connect to and work with DuckDB databases. The client provides support for connecting to DuckDB via the R database interface, as well as `dbplyr` integration, meaning that R users can query DuckDB databases using the powerful and popular `dplyr` interface, as an alternative to SQL. Along with DuckDB's core features, this makes DuckDB a powerful component of an R analytical toolchain. We cover all this in *Chapter 9*, where we go on a tour of the DuckDB R client.

- **The DuckDB Wasm client** is a full DuckDB client that has been compiled to run on Wasm, which is a **virtual machine (VM)** that runs on all modern browsers. With DuckDB Wasm, you can build web apps in JavaScript that can interact with DuckDB databases, running completely on client devices. This opens up a range of compelling possibilities for building lean analytical data apps with low-latency response times. In *Chapter 12*'s *DuckDB-powered data apps* section, we'll unpack these possibilities further, identifying contexts where you might want to consider adopting DuckDB Wasm for building data apps. We'll also cover using the DuckDB Web shell, a DuckDB CLI that runs completely within the browser, which you can try out online: `https://shell.duckdb.org`.

- DuckDB provides clients for both the **Open Database Connectivity (ODBC)** API and the **Java Database Connectivity (JDBC)** API. These are both important workhorses in the data ecosystem, being frequently used to connect analytical data applications, such as BI tools, to databases for querying. We discuss both these DuckDB integration targets in the *DuckDB integration* section in *Chapter 12*.

- **The DuckDB Arrow Database Connectivity (ADBC) client** provides an implementation of the ADBC API. This plays an analogous role to the JDBC and ODBC clients, enabling applications to connect to DuckDB databases as a data source, with the notable difference that the ADBC client makes use of Apache Arrow, an in-memory columnar data format. This is a much better fit for analytical applications, addressing the inefficiencies associated with the use of the OLTP-oriented JDBC and ODBC APIs. We'll discuss the DuckDB ADBC client in *Chapter 12*'s *DuckDB integration* section.

Next, we'll get the DuckDB CLI installed so that we're ready to dive into some hands-on DuckDB examples.

Installing the DuckDB CLI

The DuckDB CLI is made available for Windows, macOS, and Linux operating systems. For up-to-date installation options and instructions, go to the DuckDB installation page and ensure that you have the appropriate operating system for your machine selected: https://duckdb.org/docs/installation. You can choose between using a package manager to install DuckDB onto your system, or directly downloading a ZIP file that contains the DuckDB CLI executable. If you haven't used the package manager indicated in the instructions for your operating system, we suggest you take the direct download approach. Please proceed now to either install DuckDB using the package manager or download the DuckDB CLI ZIP file. If downloading directly, once the file has downloaded, unzip it and extract the DuckDB executable file into the directory you plan to work in.

> **Which version of DuckDB?**
>
> In this book, the output you'll see comes from the most recent version of DuckDB available at the time of writing (April 2024), which is 0.10.2. DuckDB Labs has indicated that the intention is for no new features to be added between this release and the 1.0.0 release, with the focus being on stability and robustness. We therefore recommend that you use the most recent version of DuckDB to work through these exercises. You may see some minor differences in the output of commands and error messages with later versions; however, this is unlikely to interfere with the exercises. If you do come across any unexpected behavior, as always, the DuckDB documentation should be considered the source of truth: https://duckdb.org/docs. We will endeavor to keep the code in the GitHub repository for this book up to date in the event that there are breaking changes: https://github.com/PacktPublishing/Getting-Started-with-DuckDB.

Starting the DuckDB CLI

Since we're working with a CLI, you'll need to open a terminal application. You may have a preferred terminal application you already use, or you can simply use the one that comes with your operating system. Note that for Windows, you can use either PowerShell or the cmd.exe application. We'll refer to the terminal application you've chosen as your *terminal* from here on.

Start by loading up your terminal and changing directory to the one you plan to work in. If you downloaded the DuckDB executable, this should be in the same directory as the one you placed the unzipped DuckDB executable in. Then, run one of the following commands appropriate for your context to start the DuckDB CLI.

Run the following command if you installed DuckDB using a package manager or if you are using Windows' cmd.exe application for your terminal application and downloaded the executable:

```
$ duckdb
```

If you are using PowerShell on Windows and downloaded the executable, run the following command:

```
$ .\duckdb
```

For macOS and Linux terminal applications and where you downloaded the executable directly, run this command:

```
$ ./duckdb
```

Note which alternative you used, as you may need to adapt subsequent duckdb executable invocations accordingly. Your terminal should now look something like this:

```
v0.10.2 1601d94f94
Enter ".help" for usage hints.
Connected to a transient in-memory database.
Use ".open FILENAME" to reopen on a persistent database.
D
```

This indicates that you're now inside the DuckDB shell, which gives you an interactive **read-eval-print loop (REPL)**, ready for you to start entering commands and interacting with DuckDB.

Working with the DuckDB CLI

Within the DuckDB shell, the D symbol indicates a waiting prompt, meaning that it's ready to accept input. In the next section, we'll go through some SQL basics and run through several SQL statements for inserting and querying data. For now, let's run a very simple query that retrieves a made-up record of values that we have specified within the query:

```
SELECT 'hello' AS greeting, 'world' AS name;
```

This query produces the following output:

```
┌──────────┬─────────┐
│ greeting │  name   │
│ varchar  │ varchar │
├──────────┼─────────┤
│ hello    │ world   │
└──────────┴─────────┘
```

After running the query, you will immediately see the resulting record, rendered as a table with a single row and corresponding column headers that indicate both the name and data type of each column.

You may have noticed that when we opened the shell, DuckDB informed us that it was connected to a transient in-memory database. This means that when you close the shell, the database being used by DuckDB, along with any data it contains, won't be persisted anywhere. For many applications, this is perfectly fine. Many ad hoc data analysis or transformation activities can be performed within a single session, with the final results being written to disk in an appropriate file format. As you continue exploring DuckDB use cases, you will discover contexts where it will be useful to persist cleaned and prepared tables to disk so that they can be reused across subsequent sessions. For these contexts, DuckDB supports opening a database as a persistent disk-based file. We can achieve this in two ways using the DuckDB CLI.

One is via the command line, by passing the path to an existing DuckDB database file as an argument to the `duckdb` executable when you load the DuckDB CLI. To try this out, first, make sure you exit any running DuckDB CLI shell by executing the `.quit` command. Then enter the following command in your terminal:

```
$ ./duckdb my_db.duckdb
```

The other way to open a persistent file-based database is via an already running DuckDB CLI shell, using the `.open` CLI command followed by the path to your desired database file:

```
.open my_db.duckdb
```

In both cases, this will result in an open connection to a disk-based DuckDB database stored in the `my_db.duckdb` file. If this file already exists on disk, DuckDB will load and start using it. If the file does not already exist, DuckDB will first create it before connecting to it. When you close the shell, any outstanding changes to the database will be written safely to the file.

The `.open` command is just one of a number of special dot commands available in DuckDB's shell. For example, the `.mode` command changes the formatting used to display tables that are returned after running a query. This is useful for quickly copying and pasting query results in other formats, such as CSV and JSON. For example, let's render the output of the query we ran previously in an HTML table. To do this, first change the output mode to `html` and then rerun the query:

```
.mode html
SELECT 'hello' AS greeting, 'world' AS name;
```

We now see the same result but formatted as an HTML table:

```
<TR><TH>greeting</TH>
<TH>name</TH>
</TR>
<TR><TD>hello</TD>
<TD>world</TD>
</TR>
```

We can also change DuckDB's output mode at the command line. By combining this feature with DuckDB's ability to pass SQL commands to the `duckdb` executable, we can start to see a glimpse of some of the versatility DuckDB has to offer:

```
$ ./duckdb -markdown -s "SELECT 'hello' AS greeting, 'world' AS name;"
```

Running this command on your terminal results in the same output, this time formatted in markdown:

```
| greeting | name  |
|----------|-------|
| hello    | world |
```

To see the available command-line parameters the `duckdb` executable supports, we can pass it the `--help` flag:

```
$ ./duckdb --help
```

For a complete list of output modes DuckDB supports, as well as a reference for other available dot commands, see the DuckDB CLI documentation: `https://duckdb.org/docs/api/cli`.

Now that we've got the DuckDB CLI set up and have seen how to work with it, we're ready to start our adventures getting started with DuckDB. In the next section, we provide a very brief introduction to working with SQL, via the DuckDB CLI. If you're already well versed in writing SQL, you're welcome to skim through or skip this section. For those of you newer to working with SQL, or if you haven't used it for a while, this will make sure that you're across some fundamentals we'll be assuming as we dive into exploring DuckDB's features.

A short SQL primer

SQL is a programming language that is specifically designed for querying, manipulating, and analyzing data. Even though SQL was originally developed in the early 1970s, it is widely used in modern data platforms and tools, with its adoption continuing to increase. SQL is a declarative language, which means that it allows us to focus on what we want to do with data, rather than having to worry about specifying the low-level steps for how to do it. It is also a rather versatile language, being frequently used across many types of applications, from ad hoc querying of data extracts to large-scale data processing pipelines and complex data analysis and reporting.

SQL's versatility across analytical data applications, combined with its ubiquity and familiarity for many data practitioners, makes it a sensible choice for DuckDB to adopt as its primary language for managing, transforming, and querying data. It's worth noting that SQL isn't the only programmatic interface for interacting with DuckDB, as we'll see later on in our DuckDB adventures. Given that many of the exercises in this book make use of SQL, in this section, we'll go through a very short primer on some SQL fundamentals for readers who are newer to working with SQL or who haven't used it for a while.

Creating your first DuckDB table

A database table is a collection of related data, organized in a tabular format consisting of rows and columns. Each row in a table represents a distinct record of the data being stored, while each column represents a specific attribute of the data stored in the table. Tables are an essential part of a database, providing a structured way to organize, store, and query data. Tables can also be linked to other tables through relationships, allowing for more complex data queries and analysis.

For our first example, let's imagine we need to store information about food types in a database. Let's create a table called `foods`, with columns describing attributes of each food, such as `food_name`, `calories`, and `is_healthy`. Each record in the `foods` table would represent a unique food type, with specific information about each food item stored in corresponding columns.

Let's now create our first DuckDB table. Creating a table in SQL involves specifying the table name, as well as the column name and data type for each column we want the table to have. The following SQL statement creates a simple table called `foods`, with the columns we outlined previously. Run this now in an open DuckDB CLI shell:

```
CREATE TABLE foods (
    food_name VARCHAR PRIMARY KEY,
    color VARCHAR,
    calories INT,
    is_healthy BOOLEAN
);
```

Note that DuckDB allows you to write multiline commands, with a semicolon (`;`) character being used to indicate the end of a SQL statement. It tells the DuckDB database engine that the current statement has ended and that it should be executed before moving on to the next statement.

You have now created a table named `foods` with the following four columns:

- `food_name`: The name of the food. We are using a VARCHAR data type, which is used to store variable-length character strings. The `food_name` column is also specified as the table's PRIMARY KEY constraint. This constraint ensures that each row in the table has a unique value for this column.

- `color`: The color of the food. This is also stored as a VARCHAR data type.

- `calories`: The calorie count of the food. This is stored as an INTEGER data type, which is used to represent whole numeric values, without any decimal places.

- `is_healthy`: An indicator of whether the food item is considered healthy. This is stored as a BOOLEAN data type, which can only take one of two values: `true` or `false`.

Once a table is created, data can be inserted into it using the INSERT statement. For example, the following SQL statement inserts a new record into the `foods` table:

```
INSERT INTO foods (food_name, color, calories, is_healthy)
VALUES ('apple', 'red', 100, true);
```

This inserts a new record with the values `'apple'` for the `food_name` column, `'red'` for the color column, `100` for the `calories` column, and `true` for the `is_healthy` column.

We can use the INSERT statement to insert multiple records at a time. The following SQL statement inserts three new records into the `foods` table:

```
INSERT INTO foods (food_name, color, calories, is_healthy)
VALUES ('banana', 'yellow', 100, true),
       ('cookie', 'brown', 200, false),
       ('chocolate', 'brown', 150, false);
```

Running this statement results in three new records being inserted into our table, bringing this up to four food items in the table. There are a range of additional features that the INSERT statement offers, which you can explore at the DuckDB documentation: https://duckdb.org/docs/sql/statements/insert.

Running your first DuckDB query

Now, let's have a look at the data we added to the foods table. To do this, we'll use the SQL SELECT command, which is used to retrieve data from one (or more) tables:

```
SELECT *
FROM foods;
```

Running this query produces the following output:

| food_name | color | calories | is_healthy |
varchar	varchar	int32	boolean
apple	red	100	true
banana	yellow	100	true
cookie	brown	200	false
chocolate	brown	150	false

Let's unpack that query, which we can see has returned the four food items that we previously inserted into the table:

- SELECT: Specifies the columns we want to retrieve from the table. We selected all columns in the target table by using the asterisk symbol (*), which functions as a wildcard. Alternatively, we could have explicitly listed one or more column names separated with commas, to return only a subset of columns.

- FROM: Specifies the name of the table we want to retrieve data from; in this case, the foods table.

As we mentioned earlier in this chapter, DuckDB's SQL dialect comes with a range of enhancements that extend traditional SQL syntax, with an eye toward a more user-friendly experience. One of these enhancements is the ability to omit the SELECT clause from a SELECT statement when returning all columns. This means that the query we just ran could be replaced with the following functionally identical and more concise query:

```
FROM foods;
```

When we created the foods table, we set a primary key on the food_name column. This instructs DuckDB to enforce the constraint that values in this column must be unique across all rows in the table. With this PRIMARY KEY constraint defined on our table, we have ensured that there is no duplication of food items in the table. We can see this constraint in action by trying to add an extra record with the name 'apple' again, but this time 'green' in color:

```
INSERT INTO foods (food_name, color, calories, is_healthy)
VALUES ('apple', 'green', 100, true);
Error: Constraint Error: Duplicate key "food_name: apple" violates
primary key constraint. If this is an unexpected constraint violation
please double check with the known index limitations section in our
documentation (https://duckdb.org/docs/sql/indexes).
```

This error indicates our insert failed, as we expected should happen. Let's check we have only the single original red apple by querying the table again. This time, we'll restrict the SELECT clause to only retrieve values from the food_name and color columns, which are the values we need to check. We'll also use a WHERE clause to apply a filter to our query so that it only returns records with a food_name value of 'apple':

```
SELECT food_name, color
FROM foods
WHERE food_name = 'apple';
```

This query returns a single result, as we expected:

food_name	color
varchar	varchar
apple	red

Now, if we did want to change the color of the existing apple record, we could achieve this by modifying the value of its color field using the UPDATE statement. The following statement shows how we can do this:

```
UPDATE foods
SET color = 'green'
WHERE food_name = 'apple';
```

Note the use of the WHERE clause to specify the condition that must be met for the rows to be updated. Because of this filter, our update is only applied to records where food_name has the value 'apple'.

Let's verify for ourselves that the existing apple record has had its color updated successfully by running the previous SELECT statement again:

```
SELECT food_name, color
FROM foods
WHERE food_name = 'apple';
```

This time, we see our green apple:

food_name	color
varchar	varchar
apple	green

Lastly, another one of DuckDB's SQL dialect enhancements is that when constructing a SELECT statement, the FROM clause can be placed before the SELECT clause. This enables queries to follow a more natural sequence, with the data source being queried placed up front, before data-source-specific components of the query, such as columns to be retrieved and column filters to be applied. Using this SQL enhancement, the query that we just ran can be rewritten as follows:

```
FROM foods
SELECT food_name, color
WHERE food_name = 'apple';
```

This brings us to the end of our condensed primer on some of the basics of SQL, which we'll be assuming you're across as we dive into exploring DuckDB's impressive range of SQL-based analytical capabilities. If you're newer to working with SQL, and you feel like you could benefit from a more comprehensive introduction to SQL to pair with this book, you could consider reading Packt Publishing's *Learn SQL Database Programming* by Josephine Bush. It's also worth noting that, in the examples we've just worked through, we've only been working with individual records, as opposed to analytical operations over columns containing many records. As we work through the book, we will explore different types of SQL-defined analytical workloads and use cases that play to DuckDB's strengths. We'll also see ways in which you can work with DuckDB using alternative non-SQL interfaces, which may appeal to data scientists and data analysts working in Python or R in particular. By the end of the book, we think that you'll see how DuckDB's adoption of SQL as its core interface enables it to be an accessible, powerful, and flexible tool for managing analytical data workloads.

Summary

In this chapter, we unpacked DuckDB, situating it within the landscape of databases and data processing tools, finding it to be a fully featured DBMS that is optimized for high performance over analytical workloads, while also being simple to install and work with by virtue of its in-process mode of operation.

We identified two broad areas of application where DuckDB is seeing much excitement and adoption: scaling and supercharging data science, data analytics, and ad hoc data-wrangling workflows, and forming a building block for operational data engineering infrastructure and interactive analytical data applications. We also outlined the properties of DuckDB that make it excel at these use cases: its performance, ease of use, versatility, powerful analytics capabilities, and an engaged community. Understanding DuckDB's strengths and capabilities is important for you to be able to spot opportunities for adopting it in your own workflows, as well as being able to recognize when an alternative data processing approach would be more appropriate.

We then looked at DuckDB deployment options, seeing the wide range of DuckDB clients available, before getting DuckDB up and running on your own machine. We then finished with a short primer on some of the fundamentals of SQL. With these preparatory steps complete, you are now ready to dive into the hands-on DuckDB SQL examples we'll be covering across the book.

In the next chapter, we're going to dive into the topic of loading data into DuckDB, by exploring DuckDB's versatile range of data ingestion patterns across a range of data sources and data formats. This will set us up for being able to explore DuckDB's powerful analytical querying and data-wrangling capabilities.

2

Loading Data into DuckDB

DuckDB is a flexible analytical database that can handle a variety of data types and workloads. A common task when working with DuckDB is loading data from external data sources, such as **comma-separated values (CSV)**, **JavaScript Object Notation (JSON)**, and Apache Parquet files. In this chapter, we will introduce the basic concepts and methods for loading data into DuckDB and provide some examples and best practices to help you get started.

You will learn how to load data into DuckDB from external data sources, how to create tables using SQL commands, and how to load data from various sources and formats, including CSV, JSON, and Parquet files, along with exploring some of the considerations when working with compressed columnar formats. We will also use DuckDB to query and analyze a public dataset, in addition to reviewing how we can export data from DuckDB.

In this chapter, we're going to cover the following main topics:

- Loading CSV files
- Loading JSON files
- Working with Parquet files
- Exploring public datasets
- Exporting data

Technical requirements

We need to get data into DuckDB so that it can be queried, transformed, and analyzed. The files and example data for these exercises are available in the `chapter_02` folder at `https://github.com/PacktPublishing/Getting-Started-with-DuckDB/tree/main/chapter_02`.

> **Important note**
>
> In this chapter, we will be learning how to load data into (and export data from) DuckDB using local files. DuckDB can also read and write data located in **Simple Storage Service (S3)**-compatible object stores, including AWS S3, Google Cloud Storage, and Azure Blob storage. We will explore using DuckDB to interact with object stores in *Chapter 5*.

Loading CSV files

CSV files are ubiquitous in the world of analytical data, which is why DuckDB comes with a powerful and flexible built-in CSV parser. The appeal of this simple text-based format is that CSV files are easy to inspect, and their tabular format is readily comprehended. While they are straightforward to produce and share, there are, however, often challenges when working with CSV files. Notably, they come in a wide variety of dialects and, often, non-standard variations. For example, despite their name, they sometimes use characters other than commas for delimiting each field—such as tabs, they may or may not have a header row with column names, and there are different approaches to escaping special characters, such as delimiters and quotes. When parsing a CSV file, the specific format of a CSV file can often be inferred but may need to be specified manually. Furthermore, CSV files don't contain an embedded schema, meaning that conversion from text values to appropriate data types (such as numbers and dates) must be performed at load time—again, either by inference or through manual specification.

Fortunately, DuckDB's CSV parser has powerful auto-detection capabilities and a range of options for providing manual specification of both the structure of target CSV files and the desired schema of column names and data types. It also has convenient features for identifying and handling errors that are encountered during parsing. In this section, we'll take you through some of DuckDB's features for loading CSV files into DuckDB in the appropriate shape while handling the challenges of working with CSV data.

The COPY statement

In *Chapter 1*, we saw how to add data to a table using the SQL INSERT statement. We can also add data to a table by loading (or importing) records from an external file into DuckDB. One of the easiest ways to load CSV data into a DuckDB database is to use the COPY statement. The DuckDB COPY statement allows us to load data from CSV, Parquet, and JSON files into an existing table, as well as export data from a table or query to the same set of file formats. Here, we're focusing on using it to load data; we'll return to using the COPY statement to export data to a file toward the end of this chapter. If you're familiar with PostgreSQL, the DuckDB COPY command has a near-identical syntax.

Before we can use the COPY statement to load data from a CSV file into a table in our database, we must first create a table that has our desired schema and is compatible with the contents of the source data file. If you happen to be targeting a table that already contains records, the COPY command will append records from the source file to the table.

The accompanying GitHub folder for this chapter has a CSV file named `food_no_heading.`
`csv`, which we're going to load into a DuckDB table using the `COPY` statement. The file contains the
following four rows of data:

```
apple,red,100,true
banana,yellow,100,true
cookie,brown,200,false
chocolate,brown,150,false
```

We don't necessarily know what the individual fields mean from the raw data alone, as the file does
not have a header row, and as with all CSV files, the data is stored as text without any accompanying
metadata. We'll be providing external information that we can use as a schema for these records.

In order to use the `COPY` statement to load data from our file, we'll create a destination table using
the `CREATE TABLE` statement. We'll also use its `OR REPLACE` clause, which instructs DuckDB to
first drop any existing table with the same name and create a new one:

```
CREATE OR REPLACE TABLE foods (
    food_name VARCHAR PRIMARY KEY,
    color VARCHAR,
    calories INT,
    is_healthy BOOLEAN
);
```

We have now created a `foods` table with appropriate names and data types for each column occurring
in our CSV file. We've also specified that the `food_name` column is the primary key for this table,
ensuring that values in this column must be unique across rows. Now, we're ready to use the `COPY`
statement to import the data from the CSV file into the table as follows:

```
COPY foods FROM 'foods_no_heading.csv';
```

Our records have now been loaded into the `foods` table. Let's verify the data was loaded correctly
by querying the table:

```
SELECT * FROM foods;
```

| food_name | color | calories | is_healthy |
varchar	varchar	int32	boolean
apple	red	100	true
banana	yellow	100	true
cookie	brown	200	false
chocolate	brown	150	false

We can see we have loaded four rows of data into the `foods` table. It's worth noting that even though our source data started as text, values in the `calories` column have been automatically cast into `INTEGER` values, and values in the `is_healthy` column have been cast into BOOLEAN values. This is a feature of the `COPY` command; it will match the data types in the schema of your target table when loading records from a CSV file.

COPY statement options

As we mentioned earlier, CSV files come in a range of different dialects and flavors. DuckDB's CSV reader has a selection of parameters for providing instructions on how source files should be parsed. DuckDB's CSV reader also comes with a highly tuned **CSV sniffer**, which is able to automatically infer the values of these parameters for many candidate CSV files, by analyzing (or sniffing) a sample of records in the file. If you know the structure of the CSV file in advance, you may want to explicitly set these parameters instead, to ensure source data does not deviate from the expected structure. Sometimes, you may encounter particularly non-standard CSV files that the CSV sniffer cannot correctly analyze. In these instances, you can configure the CSV reader via the parameters given to the `COPY` statement.

Continuing our theme of food products, imagine we receive a data file from a new supplier that is formatted differently from our previous CSV file. Notably, this file is a commonly occurring CSV variant that uses tab characters as field delimiters. This format is sometimes referred to as **tab-separated values** (TSV). Here are the contents of our new file, `foods_with_heading.tsv`:

```
food_name   is_healthy   color   calories
plum   TRUE   green   100
soup   TRUE   yellow   100
```

This file has the following notable differences from our previous CSV file:

- Tab-delimited fields rather than comma-delimited fields
- An explicit header line as the first line of the file
- Columns occurring in a different order than in our `foods` table

To load this tab-separated file of records into our `foods` table, we'll use the following COPY statement:

```
COPY foods (food_name, is_healthy, color, calories)
FROM 'foods_with_heading.tsv' (DELIM '\t', HEADER true);
```

Let's unpack this query. To handle the change in the ordering of the fields in the source file, we have given the COPY statement the target column names from the foods table, in the order they occur in the source file. To handle the other differences, we've provided explicit parameters to the CSV reader at the end of the COPY statement. We used the DELIM parameter to indicate that the field delimiter in the source file is a tab character (denoted by '\t'). We also used the HEADER parameter to indicate that the first line in the file is a header rather than a record of data so that it will be skipped when copying records.

Tab characters as delimiters and the occurrence of a header column are both commonly seen properties of CSV files; so, in this case, DuckDB's CSV sniffer is easily able to infer these corresponding parameters, and we could have omitted them.

The COPY statement accepts a range of other parameters that are passed to DuckDB's CSV reader. See DuckDB's CSV reader documentation for a complete list of available parameters: https://duckdb.org/docs/data/csv/overview.html.

Error handling with CSV reading

Now, imagine we have a source data file that contains some data that doesn't conform to our expected schema. This is a common situation, especially when dealing with large, real-world data files where we don't control the data generation process. The foods_error.csv file (which follows the same structure as the food_no_heading.csv file) has a row with an 'eighty' string occurring in the calories column, which is meant to contain integers:

```
pear,green,60,true
orange,orange,eighty,true
```

Let's see what happens when we try to load this CSV file as we did previously:

```
COPY foods FROM 'foods_error.csv';
```

On running this statement, we see that DuckDB gives us the following error:

```
Conversion Error: CSV Error on Line: 2
Original Line: orange,orange,eighty,true
Error when converting column "calories". Could not convert string
"eighty" to 'INTEGER'

Column calories is being converted as type INTEGER
```

This error indicates a type mismatch has occurred relative to the schema of the `foods` table, as DuckDB cannot convert the `'eighty'` string into an integer. The entire data load is rejected, even though there was only a single error in the file. Note that you may see a slightly different presentation of this error message in a more recent version of DuckDB. You might also see further diagnostic output that we have not shown, which indicates the CSV-parsing parameter values that were used, including which ones were automatically inferred.

This behavior may be undesirable when loading a large CSV file where we have tolerance for dropping some records that have data quality issues. We can instruct the CSV reader to skip over invalid rows in source CSV files, retaining only records that match our target schema, by setting the IGNORE_ ERRORS parameter to `true`:

```
COPY foods FROM 'foods_error.csv' (IGNORE_ERRORS true);
```

On running this statement, our file has now been loaded, with the incorrectly formatted row being ignored.

Sometimes, it's important to log information about parsing errors for later inspection. The REJECT_ SCANS and REJECT_ERRORS parameters instruct the CSV reader to store diagnostic information in persistent tables. See the documentation for more details: `https://duckdb.org/docs/ data/csv/reading_faulty_csv_files`.

File loading with read_csv

We have been using the `COPY` statement to load the contents of a CSV file into an existing table that matches the target schema. For this approach to work, we need to already know the target schema we want to load the contents of a CSV file into. In many contexts, however, we don't know what's in our CSV file or what steps we need to take to load it appropriately. This can happen when receiving a new CSV file, as we may need to start by exploring the contents of the file and experimenting with how it should be loaded. There are also situations where we might not want to load our CSV file's contents into a table, such as when transforming a CSV file directly into another cleaned CSV file or exporting to another format such as JSON or Parquet.

In such contexts, you will want to use DuckDB's `read_csv` function. This offers much more flexibility than using the `COPY` statement as it does not require you to know the schema of your CSV file upfront, allowing you to take further advantage of DuckDB's CSV sniffer, which can automatically infer appropriate data types from the file contents. It's important to note that both the `read_csv` function and the `COPY` statement use DuckDB's CSV reader under the hood, meaning that they both accept the same set of configuration parameters, some of which we've already seen.

Let's see the `read_csv` function in action with a new CSV file, `food_prices.csv`. This file has a `food_name` column containing string values and a `price` column that contains decimal values:

```
food_name,price
apple,5.63
banana,2.22
```

```
cookie,3.11
chocolate,5.21
```

Here's how we can load this file with the `read_csv` function:

```
SELECT *
FROM read_csv('food_prices.csv');
```

| food_name | price |
varchar	double
apple	5.63
banana	2.22
cookie	3.11
chocolate	5.21

This nicely structured set of query results shows us that, in addition to being able to automatically infer the structure of the CSV file, DuckDB's CSV sniffer has used the header row to name the columns, and it has inferred appropriate data types for the columns. This is particularly useful when exploring a new CSV file for the first time, as we can lean on DuckDB to do a lot of the heavy lifting of analyzing the contents of the file, helping accelerate the process of understanding its contents and getting started working with it. Sometimes it can be useful to inspect the results of the CSV sniffer to understand what configuration it has automatically inferred to pass to the CSV reader. We can do this using DuckDB's `sniff_csv` function:

```
SELECT Delimiter, HasHeader, Columns
FROM sniff_csv('food_prices.csv');
```

This shows us the field delimiter character it has detected, whether a header row was detected, and the schema that it has inferred for the columns:

Delimiter varchar	HasHeader boolean	Columns varchar
,	true	{'food_name': 'VARCHAR', 'price': 'DOUBLE'}

Figure 2.1 – A selection of columns returned by the sniff_csv function

You may like to play around with this query and inspect the other diagnostic information that the CSV sniffer returns in other columns. The Prompt column is notable for giving you back a complete read_csv query with all parameters filled with inferred values, which can be used to create an explicit query that does not rely on the CSV sniffer. It's also worth noting that you can disable the use of the CSV sniffer for automatically detecting these parameters by setting the auto_detect parameter of the read_csv function to false. This does mean, however, that you must manually specify the target schema using the columns parameter, which we'll see how to do shortly.

Next, we'll see how we can create tables from the results of the read_csv function, which allows us to take advantage of the CSV sniffer to create tables and load CSV file contents into them without having to specify the schema of tables manually.

Table creation with read_csv

As we have just seen, the read_csv function is particularly well suited for when we don't know the schema of a source CSV file in advance, as it does not have the constraint of needing an existing table with an appropriate schema that the COPY statement has. When you do need to load a CSV file's contents into a DuckDB table, it would be good if we could still leverage the CSV reader's convenient auto-detection features so that we don't need to go through the process of first identifying an appropriate schema and then explicitly creating a table with this schema. Here, we'll see how we can leverage the read_csv function to precisely do this.

To illustrate this technique, we'll continue working with the food_prices.csv file. Let's say we needed to create a table from this CSV file that contains only low-cost foods that have a price of less than $4.00.

We'll use a CREATE TABLE statement, which we'll provide with a query whose results will be used to both define the schema and populate a table called low_cost_foods. Our query will use the read_csv function, allowing us to take advantage of the CSV sniffer to infer the schema we saw in our previous example, as well as specify the price filter. Here's our query:

```
CREATE OR REPLACE TABLE low_cost_foods AS
SELECT *
FROM read_csv('food_prices.csv')
WHERE price < 4.00;
```

Note the addition of the AS keyword after the CREATE TABLE statement, which instructs DuckDB to both create the table we specified and populate it with the results of the query that occurs after it. The query is the same as we saw in the previous example, with the addition of a WHERE clause, which adds the price filter to the data selected for inclusion in the table. Looking at the contents of the low_cost_foods table we just created, we can see that it has both the expected schema and filtered food items:

```
SELECT *
FROM low_cost_foods;
```

| food_name | price |
varchar	double
banana	2.22
cookie	3.11

Using this technique allowed us to read a CSV file, detect its dialect and schema, apply a filter, and create a table with matching schema—all in a single concise SQL statement. When exploring a new CSV file that you need to load into DuckDB, this approach of using a read_csv query to first confirm that it is being parsed appropriately and then using it again to import the file's contents into a new table is a particularly convenient pattern.

Date formats

So far, we have been able to arrive at the correct types of data loaded from CSV files by relying on the CSV reader's type auto-detection or by leveraging the schema of a target table with the COPY command. Sometimes, we need to give DuckDB a helping hand to understand the format of data during import. Dates and times can be especially tricky, as they can come in a wide variety of string-based formats. DuckDB can automatically detect several common string representations of dates and times; however, sometimes you may find that you need to explicitly define the format present in the source column in order to load it as a DATE or TIMESTAMP type.

Suppose we have a CSV file named food_orders.csv that contains the following contents:

```
food_name,order_date,quantity
cake,20231225,12
sushi,20230126,21
```

Note the `order_date` column is specified in the YYYYMMDD format, meaning the `20231225` value represents the date December 25, 2023. Let's see what happens when we attempt to read this file using the `read_csv` function:

```
SELECT *
FROM read_csv('food_orders.csv');
```

| food_name | order_date | quantity |
varchar	int64	int64
cake	20231225	12
sushi	20230126	21

We can see that the `order_date` column has been detected as having the type of INTEGER rather than the desired DATE type. The date format found in this file is a less common one that DuckDB's auto-detection feature does not recognize. This means we'll have to instruct the CSV reader how it should parse it. To do this, we need to pass two parameters to the `read_csv` function:

- `dateformat`: A string that specifies the date format used to parse dates.
- `types`: An explicit mapping of column names to data types that override the identified types, in the form of a STRUCT value.

The `dateformat` parameter accepts strings that use DuckDB's date formatting specification to define a target date format. The value that represents the date format in our CSV file is `'%Y%m%d'`. Additionally, we also need to tell the CSV reader that we want the `order_date` parameter to be read as a DATE type using our specified date format. This is what we'll use the `types` parameter for. Putting these together, here's our new query:

```
SELECT *
FROM read_csv(
    'food_orders.csv',
    dateformat='%Y%m%d',
    types={'order_date': 'DATE'}
);
```

| food_name | order_date | quantity |
varchar	date	int64
cake	2023-12-25	12
sushi	2023-01-26	21

We can see that we've successfully loaded our CSV file with the `order_date` column as a DATE type. If you find yourself needing to read text data files containing non-standard date formats, you will want to consult the DuckDB documentation for the complete date formatting specification: `https://duckdb.org/docs/sql/functions/dateformat`.

The `types` parameter is convenient when you need to override a subset of the column types that the CSV reader would have otherwise assigned. Sometimes, you might need to take more control over the reading of CSV files by specifying the full schema across all columns. To do this, you use the `columns` parameter, which is passed a STRUCT value containing every column name you want to read from the CSV file, along with a corresponding data type for each column. Let's use this approach to achieve the same result as in our previous example:

```
SELECT *
FROM read_csv(
    'food_orders.csv',
    dateformat='%Y%m%d',
    columns={
        'food_name': 'VARCHAR',
        'order_date': 'DATE',
        'quantity': 'INTEGER'
    }
);
```

This gives us the same outcome as before, with the `order_date` column being loaded as a DATE type but through a fully specified schema, where we explicitly identified each column and its target data type. For our needs here, using the `types` parameter is much more convenient as we only need to override a single type. However, if you ever need to specify the types of multiple columns, while also optionally reading a subset of columns from a source CSV file, the `columns` parameter allows you a greater level of control to achieve this.

When to use COPY versus read_csv

We've now seen two ways we can create tables from the records in a CSV file:

- Using the COPY statement to copy records from a CSV file into an existing table
- Using the `read_csv` function in conjunction with the CREATE TABLE statement to read records and create a table

You may have found yourself wondering which strategy you should use. The answer is that it depends on the context of what you're trying to do. If you already have an existing table that you want to import records into that matches the structure of your CSV file (or can readily create one), then using COPY is preferred, as DuckDB's CSV reader will use the target table's existing schema rather than attempting to infer one from the CSV file. This will make the CSV reading faster, and it also means that you get data validation for free, eliminating errors from incorrect data types being inferred.

It's also worth noting that the COPY statement will append records from your CSV file to the end of the table after any existing records, which is another feature that may influence your choice of loading strategy.

In situations where you don't have an existing table whose schema matches the destination table or are assembling this data to do something other than table creation, then using the read_csv function is more likely the way to go. This is more frequently the context you'll find yourself in when performing ad hoc analysis or **exploratory data analysis (EDA)** over new datasets.

We have seen how to load individual data files into DuckDB. Next, we'll see how we can use DuckDB to load multiple files into a destination table.

Loading multiple files

We can instruct read_csv to load multiple files by using a file mask. A file mask, also known as a glob pattern, is a pattern that specifies a set of files or directories based on certain criteria. For example, *.csv matches all files that end with .csv, and /home/kim/pizza* matches all files under the /home/kim directory starting with the pizza prefix.

For the next example, we have three pizza-related CSV files in a directory called food_collection. As they all share a common format, we can load them all with DuckDB using a single read_csv call with a file mask of pizza*.csv. We can also use the filename=true parameter to instruct DuckDB to extend the result schema with an additional filename column, which contains the path of the source file that each record originates from. This is useful when trying to locate the origin of a row of data when processing a dataset spread across many files:

```
SELECT food_name, color, filename
FROM read_csv(
    'food_collection/pizza*.csv',
    filename=true
);
```

food_name varchar	color varchar	filename varchar
Margherita	mixed	food_collection/pizza_1.csv
Chicago	yellow	food_collection/pizza_1.csv
Pepperoni	red	food_collection/pizza_1.csv
Hawaiian	yellow	food_collection/pizza_2.csv
Neapolitan	red	food_collection/pizza_2.csv
BBQ	red	food_collection/pizza_3.csv
Veggie	mixed	food_collection/pizza_3.csv

We can see that we have successfully loaded our records from across all three CSV files and that each record contains a `filename` field indicating the path of its originating source file. This illustrates how using file masks is a great time-saver when there are numerous files of the same type and schema that need to be loaded together.

Mixed schemas

We sometimes face situations where we wish to load many files, but they don't all follow the same schema. Take, for example, a situation where we wish to load a collection of food CSV files, but the number and order of columns are different across files.

Our pizza files have four columns (`food_name`, `color`, `calories`, and `is_healthy`), and our fast-food files have three columns (`food_name`, `color`, and `calories`). Let's introduce a new set of files that contain types of Asian food; these files have four columns but in different order (`food_name`, `calories`, `is_healthy`, and `color`).

We want to load all files found in the `food_collection` directory, regardless of how they are organized. However, when we load these multiple files into DuckDB using `read_csv`, we hit a problem:

```
SELECT *
FROM read_csv('food_collection/*.csv');

Invalid Input Error: Mismatch between the schema of different files
```

DuckDB has helpfully informed us that there was an error due to a schema mismatch. This is to be expected as, by default, DuckDB will read the schema from the first file it processes and then unify columns in the following files by position. For this to succeed, all files must have the same schema.

Fortunately, the `read_csv` function provides the handy `union_by_name` option, which allows us to union records across different files by name rather than by position. This allows us to load and combine records across multiple files with different sets of column names, provided that the same column names across different files have the same type:

```
SELECT *
FROM read_csv(
    'food_collection/*.csv',
    union_by_name=true
);
```

| food_name | calories | is_healthy | color |
varchar	int64	boolean	varchar
sushi	60	true	mixed
pho	70	true	yellow
burger	60		mixed
fries	35		yellow
...			
Hawaiian	60	false	yellow
Neapolitan	45	false	red
BBQ	62	false	red
Veggie	32	true	mixed
salad	50	true	green
yogurt	20	true	white

We can see that our query has successfully loaded the fields from across the CSV files into appropriate target columns, even though they occur in different positions within the files. Note that in the case of the `is_healthy` column, which does not occur in our fast-food CSV files, records coming from these files have NULL values for this column (represented in the output as empty cells). This is the expected behavior when performing a union by name, matching the semantics of DuckDB's UNION ALL BY NAME SQL operation. The NULL value indicates that no value was found while combing the files.

The `union_by_name` feature is a powerful way to align mismatched schemas, addressing a common challenge when loading data across multiple files.

Loading JSON files

JSON is a popular and open file format for storing and exchanging data and has good integration with many programming languages, libraries, and data systems. A JSON object is written inside curly braces and can contain multiple name-value pairs. A name-value pair consists of an attribute name in double quotes, followed by a colon, followed by a value. JSON values can include strings, numbers, and Boolean data types, as well as nested objects and array data types, which are represented as comma-separated sequences of values, wrapped with square braces. Here's an example of a simple JSON object:

```
{"food_name":"Hawaiian Pizza", "quantity":6}
```

DuckDB can read JSON files with the `read_json` function in a variety of formats. In the wild, you may encounter a range of different types of JSON files, with records represented using different conventions. Two common flavors of JSON data that you may encounter in the wild are newline-delimited JSON objects and arrays of JSON objects. DuckDB supports reading JSON records represented in both these styles.

JSON Lines – Newline Delimited JSON

Take the situation where we have JSON records where each record is separated by a newline character. **JSON Lines** (or **NDJSON**, short for **Newline Delimited JSON**) is a common format for data exchange, especially in the context of streaming data systems. It's characterized by each record being represented as a JSON object, occurring on its own line, which is to say, each record is separated by a newline character. Let's imagine we have a food distribution system that produces newline-delimited JSON records. The `pizza_orders_records.json` file contains two JSON records describing pizza orders formatted as newline-delimited JSON data:

```
{"food_name":"Hawaiian Pizza", "quantity":6,
 "order_date":"2023-02-24 15:00:00"}
{"food_name":"Vegetarian Pizza", "quantity":2,
 "order_date":"2023-03-25 15:00:00"}
```

DuckDB's read_json function takes a format parameter, which tells the JSON parser what shape to expect records as. The string value that indicates a file should be parsed as newline-delimited JSON is 'newline_delimited'. Here's a query using read_json that will load our records from this file:

```
SELECT *
FROM read_json(
    'pizza_orders_records.json',
    format='newline_delimited'
);
```

food_name varchar	quantity int64	order_date timestamp
Hawaiian Pizza	6	2023-02-24 15:00:00
Vegetarian Pizza	2	2023-03-25 15:00:00

Note that DuckDB's parser has automatically inferred appropriate data types for the query results. In particular, the order_date column has been correctly assigned a TIMESTAMP field. We should also point out that DuckDB's JSON parser can usually detect the structure of a JSON file and infer the appropriate value of the format parameter. We could have omitted the format parameter when reading this file, but we have included it explicitly as an illustrative example, as there are times when you may need or want to ensure that read_json is reading a file as newline-delimited JSON.

JSON array objects

JSON data files can also be arranged as an array of objects. A JSON array is written inside square brackets and can contain multiple values or objects, as in this example:

```
[
    {
        "food_name": "Margherita Pizza",
        "quantity": 12,
        "order_date": "2023-01-20T15:00:00Z"
    },
    {
        "food_name": "Pepperoni Pizza",
        "quantity": 22,
        "order_date": "2023-01-22T15:00:00Z"
    }
]
```

To load an array of JSON objects, the `format` parameter of the `read_json` function should have the `'array'` value:

```
SELECT *
FROM read_json(
    'pizza_orders_array_of_records.json',
    format='array'
);
```

food_name varchar	quantity int64	order_date timestamp
Margherita Pizza	12	2023-01-20 15:00:00
Pepperoni Pizza	22	2023-01-22 15:00:00

DuckDB's JSON parser has again inferred the data types and loaded the records nested in the JSON array. Note that just as in the newline-delimited example, we could have omitted the `format` parameter to parse this file, but we have included it to illustrate this parameter's function. We have only scratched the surface of the powerful JSON features available in DuckDB. In *Chapter 6*, we'll go deeper into DuckDB's JSON parsing capabilities, as well as how to work with and model JSON data in DuckDB.

Working with Parquet files

Apache Parquet is an open source file format that is designed for efficient storage and retrieval of data. Their columnar-oriented format combined with the use of compression to reduce storage space and I/O cost of reading and writing make these files well suited for storing and retrieving large amounts of structured and semi-structured data for analytical applications.

Parquet files are encoded in a binary format, so you cannot view them as text files as you might with a CSV file. Parquet files are self-describing in that each file contains both data and metadata describing the schema of the data within the file. This means that column names, their data types, and summary information about the number of rows and columns are encoded within the file. This contrasts with CSV and JSON files, which contain purely text data without an embedded schema. In addition to performance gains, this is one of the notable benefits of Parquet files, as their built-in schema greatly reduces the chance of data integrity errors that can occur when having to explicitly convert source data fields into appropriate data types.

Parquet files are especially useful for scenarios where data consumption is read-intensive and analytical in nature, such as data warehousing, business intelligence, and machine learning. Parquet files are frequently used by popular data tools and frameworks in the big data ecosystem, such as Apache Spark, Apache Hive, Apache Impala, Presto, and Trino. DuckDB can read and write Parquet files as well as leverage their columnar format and rich metadata to optimize data retrieval and processing. This unlocks highly efficient analytical workflows, as well as making DuckDB a valuable building block for analytical data infrastructure.

Inside the accompanying files for this chapter is a Parquet file called food_orders.parquet. Before we load it into DuckDB, let's use the DuckDB parquet_schema function to view the internal schema contained within this Parquet file:

```
SELECT name, type, converted_type
FROM parquet_schema('food_orders.parquet');
```

| name | type | converted_type |
varchar	varchar	varchar
duckdb_schema		
food_name	BYTE_ARRAY	UTF8
order_date	INT32	DATE
quantity	INT32	INT_32

This shows us Parquet-specific metadata about the columns in the file, so we don't necessarily see DuckDB data types occurring in this output. However, the converted_type field gives us enough information to infer how DuckDB will load these columns:

- food_name: UTF8 indicates a Unicode string encoding, so this corresponds to a DuckDB VARCHAR type.

- order_date: As you might expect, DATE corresponds to a DuckDB DATE type.

- quantity: INT_32 corresponds to a DuckDB INTEGER type.

Now, let's query the contents of this Parquet file using DuckDB's read_parquet function, which will read the file and consult the internal metadata to assign appropriate DuckDB types:

```
SELECT *
FROM read_parquet('food_orders.parquet');
```

| food_name | order_date | quantity |
varchar	date	int32
apple	2023-12-25	10
cookie	2023-01-26	25
chocolate	2023-04-01	60

It's worth highlighting again that we didn't need to specify a schema in order to read this file appropriately, as DuckDB was able to use the embedded schema in the Parquet file. This contrasts with what happened when we queried the text-based CSV and JSON files, where DuckDB used its automatic detection features to infer an appropriate schema from a sample of records in the file; furthermore, we had to provide explicit instructions to load one of the fields as a DATE type. Here, no additional parsing instructions were required to load the order_date column as a DATE type, as it was stored as a Parquet DATE type already.

Lastly, if we want to quickly see the DuckDB data types that would be loaded from a Parquet file, we can get DuckDB to give us this information directly by applying the DESCRIBE statement to a read_parquet query:

```
DESCRIBE
SELECT *
FROM read_parquet('food_orders.parquet');
```

This gives us the following information:

column_name varchar	column_type varchar	null varchar	key varchar	default varchar	extra varchar
food_name order_date quantity	VARCHAR DATE INTEGER	YES YES YES			

Figure 2.2 – The resulting DuckDB schema from reading a Parquet
file, as shown by the DESCRIBE statement

> **Best practice when persisting transformed data as files**
>
> It is good to get in the habit, where possible, of using Parquet for saving transformed data to disk, rather than CSV or JSON. Parquet files are very efficient for storage, providing a compact way of storing large amounts of data with smaller footprints. Furthermore, persisting the schema alongside the data will make your workflows and pipelines more robust. These properties make Parquet files a superior format for most use cases than text-based files such as CSV and JSON. Unless you specifically require CSV or JSON format for data interchange, when storing transformed and cleaned data in a file format, we recommend persisting data as Parquet files.

The self-describing nature of Parquet files has the very compelling benefit of eliminating any ambiguity in how to interpret column values, drastically reducing the likelihood of errors resulting from type mismatches, as well as eliminating the need for maintaining dedicated ingestion code required for converting textual representations into the target type.

Exploring public datasets

Public datasets are collections of data that are made available to the public by various sources, such as governments, organizations, and researchers. Public datasets can be useful for analyzing trends and patterns in different domains, such as health, education, environment, or social impact.

DuckDB is a powerful tool for exploring, understanding, and gaining insights from public datasets. In this section, we'll work with a public dataset that has been made available in CSV format. We'll use DuckDB to load it in an appropriate form, summarize it, and export it to another format. This worked example will allow us to showcase some of DuckDB's versatile features that make it well suited for analytical workflows.

Bike-share station readings

We are going to be exploring the *Melbourne Bike Share* dataset, which provides historical data from the Melbourne Bike Share service, which operated from 2010 to 2019, and is made available by Melbourne Data, the City of Melbourne's open data platform (`https://data.melbourne.vic.gov.au`). The dataset contains the locations of bike-share pods and sensor readings corresponding to the number of bikes located in each pod at different times. Collected over 6 years from docking sensor readings, this dataset enables the exploration of historical usage patterns of the Melbourne Bike Share program.

To download the CSV file containing this dataset, follow these steps:

1. Visit the dataset's page in the data portal: `https://data.melbourne.vic.gov.au/explore/dataset/melbourne-bike-share-station-readings-2011-2017/information/`.

2. Download the compressed CSV file locally to your local machine.

3. Unzip the downloaded file and extract its contents.

You should have a file called `74id-aqj9.csv`. Once unzipped, this file should be roughly 943 MB in size and will contain around 5.6 million rows.

Loading the bike station readings

As the data is in CSV format and we don't yet know what the target schema will be, we'll start by using the `read_csv` function to inspect some of the data. We don't need to retrieve all 5 million rows for this, so let's use a `LIMIT` clause to look at 5 records:

```
SELECT *
FROM read_csv('74id-aqj9.csv')
LIMIT 5;
```

We get the result as follows:

ID int64	NAME varchar	TERMINALNAME int64	NBBIKES int64	…	INSTALLDATE int64	LAT double	LONG double	LOCATION varchar
2	Harbour Town - Doc…	60000	10	…	1313724600000	-37.814022	144.939521	(-37.814022, 144.9…
4	Federation Square …	60001	9	…		-37.817523	144.967814	(-37.817523, 144.9…
6	State Library - Sw…	60003	1	…		-37.810702	144.964417	(-37.810702, 144.9…
7	Bourke Street Mall…	60004	4	…		-37.813088	144.967437	(-37.813088, 144.9…
8	Melbourne Uni - Ti…	60005	8	…		-37.79625	144.960858	(-37.79625, 144.96…

5 rows   16 columns (8 shown)

Figure 2.3 – Five records queried from the bike-share dataset

We can see that the CSV reader has identified and loaded 16 columns; however, because there are so many columns, some have been omitted in order to fit into the terminal window. You may see a different number of columns truncated, depending on the width of your terminal. We'd like to get a sense of the schema that was inferred from this CSV file, so let's use the DESCRIBE statement:

```
DESCRIBE
SELECT *
FROM read_csv('74id-aqj9.csv');
```

This shows us the column names and data types identified by the CSV sniffer when loading our CSV file:

column_name varchar	column_type varchar	null varchar	key varchar	default varchar	extra varchar
ID	BIGINT	YES			
NAME	VARCHAR	YES			
TERMINALNAME	BIGINT	YES			
NBBIKES	BIGINT	YES			
NBEMPTYDOCKS	BIGINT	YES			
RUNDATE	BIGINT	YES			
INSTALLED	BOOLEAN	YES			
TEMPORARY	BOOLEAN	YES			
LOCKED	BOOLEAN	YES			
LASTCOMMWITHSERVER	BIGINT	YES			
LATESTUPDATETIME	BIGINT	YES			
REMOVALDATE	VARCHAR	YES			
INSTALLDATE	BIGINT	YES			
LAT	DOUBLE	YES			
LONG	DOUBLE	YES			
LOCATION	VARCHAR	YES			

16 rows   6 columns

Figure 2.4 – The schema inferred by the CSV sniffer for the bike-share CSV file

From the inferred schema, we can see some possible issues with columns that may not have loaded with appropriate data types. The RUNDATE column, for example, sounds like it may contain a date, but it has been identified as a BIGINT numerical type. This schema information, however, doesn't show us any of the records loaded from the file, which we'll need to diagnose the issue.

To better understand the contents of the CSV file, we'll use the .mode command to alter the output format that the DuckDB CLI uses for displaying tables. This is an example of the DuckDB CLI's dot commands, which we touched on briefly in our initial exploration of the DuckDB CLI back in *Chapter 1*. Dot commands are unrelated to the SQL language and provide control over DuckDB's command-line experience. You can see the full range of supported dot commands in the DuckDB documentation: https://duckdb.org/docs/api/cli/dot_commands.

The following dot command will update our DuckDB CLI session to use the line output format, which changes how tables are rendered by displaying each column's value on a new line:

```
.mode line
```

This will allow us to view every column value in each row when we run our previous query, this time limiting it to a single row:

```
SELECT *
FROM read_csv('74id-aqj9.csv')
LIMIT 1;
```

The following is the result:

```
                ID = 2
              NAME = Harbour Town - Docklands Dve - Docklands
      TERMINALNAME = 60000
           NBBIKES = 10
      NBEMPTYDOCKS = 11
           RUNDATE = 20170422134506
         INSTALLED = true
         TEMPORARY = false
            LOCKED = false
LASTCOMMWITHSERVER = 1492832566010
  LATESTUPDATETIME = 1492832565029
       REMOVALDATE =
       INSTALLDATE = 1313724600000
               LAT = -37.814022
              LONG = 144.939521
          LOCATION = (-37.814022, 144.939521)
```

Figure 2.5 – A single row, displayed using the CLI's Lines output format

We can see that each column is now presented on a separate line, allowing us to see a complete row of data. Before we continue, we'll revert our display mode back to the default duckbox:

```
.mode duckbox
```

Did you notice the value of the RUNDATE column for this row was 20170422134506? We know that the dataset includes data from 2017, so it's likely that the value is a timestamp, with 20170422 representing the date April 22, 2017, and 134506 representing the time 13:45:06. Our task then, is to get DuckDB to correctly load this field as a TIMESTAMP type. This is a necessary step if we want to enable time series analysis of this dataset.

Specifying date load formats

Now that we've been able to have a look at the contents of this dataset, let's load it into a DuckDB table. Just like we did earlier in this chapter when instructing the CSV reader how to load a field as a DATE type, we'll need to provide instructions on how to parse the RUNDATE field as a TIMESTAMP type. We'll use the read_csv function's types parameter again to indicate which column we want to be loaded as a TIMESTAMP type. Additionally, we'll use the timestampformat parameter to provide the string format used to encode timestamps in this column, which takes the following form: %Y%m%d%H%M%S. Let's put this together into a single statement that will create a bikes table and load into it the appropriately parsed contents of our CSV file:

```
CREATE OR REPLACE TABLE bikes AS
SELECT *
FROM read_csv(
    '74id-aqj9.csv',
    timestampformat='%Y%m%d%H%M%S',
    types={'RUNDATE': 'TIMESTAMP'}
);
```

This command may take a few seconds to complete as it needs to read and parse a CSV file with over 5 million rows.

Summarizing the bike station readings

Now we have created our bikes table, let's see how many rows we loaded into it:

```
SELECT count(*) FROM bikes;
```

```
┌─────────────┐
│ count_star()│
│    int64    │
├─────────────┤
│    5644554  │
└─────────────┘
```

That's over 5.6 million rows; not bad for a few seconds of processing time!

Beyond simply counting the rows of this dataset, DuckDB provides a handy SUMMARIZE command to quickly get a feel for the range, types, and distribution of data across a query or a table. To give you a sense of how to use this command, here's how we can summarize this dataset by applying the SUMMARIZE command to a query:

```
SUMMARIZE SELECT * FROM bikes;
```

In this case, because we want to summarize the whole contents of an existing table, the simpler (but equivalent) approach is for us to summarize the bikes table directly:

```
SUMMARIZE bikes;
```

This command produces a wealth of information about our dataset—so much, in fact, that the DuckDB CLI will likely truncate the output by omitting some columns from the middle. You should try the preceding commands for yourself so that you can see this. We'll get around this here by selecting only a subset of the columns returned by the SUMMARIZE command using a simple subquery:

```
SELECT * EXCLUDE(count, q25, q50, q75)
FROM (SUMMARIZE bikes);
```

In this query, we've placed our SUMMARIZE command in a subquery, which we've selected only a subset of columns from. To do this without having to explicitly list every column we want to include, we've used DuckDB's convenient EXCLUDE clause, which allows us to specify a much shorter list of columns we want to exclude from our results. We'll discuss this feature of DuckDB alongside other convenient DuckDB SQL enhancements in *Chapter 10*. Here's the summary information that's produced from this query:

column_name varchar	column_type varchar	min varchar	.	std varchar	null_percentage decimal(9,2)
ID	BIGINT	2	.	15.312326868433466	0.00
NAME	VARCHAR	205 Bourke St nea.	.		0.00
TERMINALNAME	BIGINT	60000	.	14.88960317933117	0.00
NBBIKES	BIGINT	0	.	6.2807318203960305	0.00
NBEMPTYDOCKS	BIGINT	0	.	6.04798799208204	0.00
RUNDATE	TIMESTAMP	2015-04-22 12:36:42	.		0.00
INSTALLED	BOOLEAN	true	.		0.00
TEMPORARY	BOOLEAN	false	.		0.00
LOCKED	BOOLEAN	false	.		0.00
LASTCOMMWITHSERVER	BIGINT	1429651366010	.	25759604591.447823	0.00
LATESTUPDATETIME	BIGINT	1429607109147	.	25758911581.78492	0.02
REMOVALDATE	VARCHAR		.		100.00
INSTALLDATE	BIGINT	1277175180000	.	52703384644.17541	22.01
LAT	DOUBLE	-37.867068	.	0.01628828854737686	0.00
LONG	DOUBLE	144.935296	.	0.012684566548855946	0.00
LOCATION	VARCHAR	(-37.79625, 144.96.	.		0.00
16 rows				8 columns (5 shown)	

Figure 2.6 – Information about the bikes table produced using the SUMMARIZE command

We can see that the SUMMARIZE command has given us information about the data types for each column, as well as their range of values, number of unique values, average value, standard deviation, and the proportion of rows containing NULL values. This is a great first step in understanding the nature of the data within our dataset.

We have taken quite a few steps to explore this bike dataset, so let's take a moment to review them. We took 5.64 million rows of data from our CSV file and loaded it into DuckDB using the read_csv function. This allowed us to identify the appropriate configuration to give to the read_csv function so that when we created a table to load this dataset into, it would have columns with the data types we wanted. Once the records from our CSV file were loaded into the table (only taking a few seconds), we used DuckDB's SUMMARIZE command to gain a high-level understanding of the properties of the dataset we now had in our DuckDB database.

Exporting data

So far, we have been importing data from a variety of data sources into DuckDB. We also often need to export data from DuckDB into a file, perhaps to transfer to another database system or to share our data with others. Let's discuss how we can achieve that in the following sections.

Exporting a table into a CSV file

We can use the COPY ... TO command to export data from DuckDB to an external CSV, JSON, or Parquet file. This command can be called either with a table name or a query and will export the corresponding data to disk in the desired file format. Let's try this out by creating a subset of our bike-share dataset and exporting it as a CSV file.

We'll first create a table called bike_readings_april containing bike rides that occurred in April:

```
CREATE OR REPLACE TABLE bike_readings_april AS
SELECT *
FROM bikes
WHERE RUNDATE BETWEEN '2017-04-01' AND '2017-04-30';
```

With the bike_readings_april table created, let's now copy its content to the bike_readings_april.csv file:

```
COPY bike_readings_april TO 'bike_readings_april.csv';
```

This command will export the entire contents of the table, creating a CSV file in the current working directory. DuckDB infers the export file format based on the supplied filename suffix. Here's a sample of records from the file we just exported:

```
ID,NAME,TERMINALNAME,NBBIKES,NBEMPTYDOCKS,RUNDATE,INSTALLED,TEMPORARY,
2,Harbour Town - Docklands Dve - Docklands,60000,10,11,2017-04-22 13:4
4,Federation Square - Flinders St / Swanston St - City,60001,9,18,2017
6,State Library - Swanston St / Little Lonsdale St - City,60003,1,10,2
7,Bourke Street Mall - 205 Bourke St - City,60004,4,7,2017-04-22 13:45
8,Melbourne Uni - Tin Alley - Carlton,60005,8,11,2017-04-22 13:45:06,t
9,RMIT - Swanston St / Franklin St - City,60006,3,4,2017-04-22 13:45:0
10,St Paul's Cathedral - Swanston St / Flinders St - City,60007,7,4,20
11,MSAC - Aughtie Dve - Albert Park,60008,12,15,2017-04-22 13:45:06,tr
```

Figure 2.7 – A sample of records from the bike_readings_april.csv file,
created by exporting the bikes table to CSV format

While the format looks quite like the original CSV file we started with, it's worth noting that DuckDB has serialized the RUNDATE column using the more standard ISO *8601* format, which has the form YYYY-MM-DD HH:MM:SS.

The sample of our exported CSV file also shows us that DuckDB defaults to writing a header row, as well as using a comma for the field delimiter. In the same way we can specify these parameters when reading a CSV file, we can also set these parameters to control how DuckDB formats our output file. If, for example, we wanted to produce a tab-delimited text file that does not have a header row, we would use the following COPY ... TO command:

```
COPY bike_readings_april TO 'bike_readings_april.tsv'
    (HEADER false, DELIM '\t');
```

This command will produce a bike_readings_april.tsv tab-delimited file. Next, we'll look at exporting data from DuckDB into JSON format.

Exporting JSON data

If you need to export data from DuckDB for use in a consuming service or another data processing tool, you may benefit from using JSON, which provides a richer and more flexible data format than CSV.

Let's imagine we want to export the contents of the `bike_readings_april` table to a JSON file. Additionally, we only want to extract the name and date of each reading, as well as prescribe the format of the date values to be of the form `22 April 2017`. We can achieve all of this with a single DuckDB `COPY ... TO` command:

```
COPY (
    SELECT NAME, RUNDATE
    FROM bike_readings_april
) TO 'bike_readings_april.json'
    (FORMAT JSON, TIMESTAMPFORMAT '%d %B %Y');
```

In this command, we used the `FORMAT` parameter to specify that we want JSON output, and we also specified the target textual representation for all dates using the `TIMESTAMPORMAT` parameter. Executing this statement will run the query and create a `bike_readings_april.json` file in the current working directory. Here are some of the records from this exported JSON file:

```
{"NAME":"Sandridge Bridge - Southbank","RUNDATE":"22 April 2017"}
{"NAME":"Beach St - Port Melbourne","RUNDATE":"22 April 2017"}
{"NAME":"New Quay Prom / Harbour Esp - Docklands","RUNDATE":"22 April 2017"}
{"NAME":"Queensbridge St / Yarra River - Southbank","RUNDATE":"22 April 2017"}
{"NAME":"Southern Cross Station - Spencer St - City","RUNDATE":"22 April 2017"}
{"NAME":"Gasworks Arts Park - Pickles St - Albert Park","RUNDATE":"22 April 2017"}
{"NAME":"Cleve Gardens - Fitzroy St - St Kilda","RUNDATE":"22 April 2017"}
```

Figure 2.8 – A sample of exported JSON records from the bike_readings_april.json file

We can see that our JSON file has been created with the two fields we asked for and that the dates have been formatted with JSON strings using our desired format.

Exporting Parquet with Hive partitioning

If we are exporting a large amount of data, it can become impractical to use a single file. We can use DuckDB partitioning to split file exports into multiple parts. **Hive partitioning** (popularized by the Apache Hive data warehouse project) is a technique to logically organize data into smaller subsets based on a partitioning rule of the data. A common approach is to split large datasets by an attribute such as the creation date of the record. We will use Hive partitioning to split the data into chunks organized by creation date.

A logical way to perform the export of the data in our `bike_readings_april` table is by splitting records into multiple files based on the day of the sensor reading found in the RUNDATE column. We will need to ignore the time component of our timestamp as we wish to co-locate records for a single day into a named day partition. For example, we wish to place all records for April 2, 2017, into a partition called 2017-04-02 regardless of the time of day the sensor reading occurred. To truncate the RUNDATE timestamp column down to the day, we will use the `date_trunc` function to truncate our timestamp with down-to-the-day precision.

Putting all this together, we can instruct DuckDB to export our `bike_readings_april` table into daily partitions under the `bike_readings` directory by using the PARTITION_BY parameter against the `rundate_truncated` field:

```
COPY (
    SELECT *,
    date_trunc('day', rundate) AS rundate_truncated
    FROM bike_readings_april
) TO 'bike_readings'(
    FORMAT PARQUET,
    PARTITION_BY (rundate_truncated),
    OVERWRITE_OR_IGNORE true
);
```

This will create a Hive-partitioned directory, `bike_readings`, containing multiple Parquet files, each stored within a subdirectory corresponding to the day of the RUNDATE partition the file lands in, as follows:

```
bike_readings
├── rundate_truncated=2017-04-01
|     └── data_0.parquet
├── rundate_truncated=2017-04-02
|     └── data_0.parquet
├── rundate_truncated=2017-04-03
|     └── data_0.parquet
├── rundate_truncated=2017-04-04
|     └── data_0.parquet
├── rundate_truncated=2017-04-05
|     └── data_0.parquet
├── rundate_truncated=2017-04-06
|     └── data_0.parquet
```

Splitting exports into multiple Parquet files can simplify storage and improve the performance of systems that need to read the data by reducing the amount of data scanned and processed. Hive partitioning is commonly used in data lake storage for efficient retrieval by distributed SQL query engines, such as Amazon Athena and Trino, which can leverage the partition keys to reduce the number of files needed to be scanned. We will revisit Hive partitioning in *Chapter 4* to read back partitioned files on disk.

Summary

In this chapter, we learned how to load data into and unload data from DuckDB. We saw data as text formats in the form of CSV and JSON, as well as the self-describing binary columnar format, Apache Parquet.

We learned techniques to format data during import and skip records with errors, and we saw how DuckDB can support a variety of changing schemas and data types. We also learned how to find, process, and summarize public datasets and saw how DuckDB can be used to export data for consumption by analytical systems.

Now that we know how to load data into DuckDB, the next chapter will cover techniques for using DuckDB for data manipulation in order to transform your data. You will learn how to clean and reshape data using SQL and use these approaches to manipulate data from different sources and formats. You will also see how to interact with data located on remote systems, such as data located on remote web servers.

3

Data Manipulation with DuckDB

This chapter introduces the basics of data manipulation with **DuckDB**. You will learn how to clean and reshape data using SQL queries and functions and manipulate data from different sources and formats. You will explore how to use aggregate functions, window functions, and common table expressions to perform complex operations on data. By the end of this chapter, you will have a solid grounding in how you can use DuckDB to clean, reshape, and generally wrangle data appropriately for your own analytical workflows.

In this chapter, we're going to cover the following main topics:

- Data wrangling
- Data import and manipulation
- Altering tables and views
- Aggregate functions and common table expressions
- Joining data from multiple tables

The techniques in this chapter will help you make your data more understandable and give you the experience to structure your DuckDB analytic pipelines. Let's dive in.

Technical requirements

You will find the code examples and data files for this chapter in the chapter_03 folder in the book's GitHub repository, which can be found at https://github.com/PacktPublishing/Getting-Started-with-DuckDB/tree/main/chapter_03.

Data wrangling with DuckDB – cleaning and reshaping data

Data wrangling is the process of transforming raw data into a more usable shape, making it more appropriate and valuable for a variety of downstream applications. When performing data wrangling, data practitioners must continually work towards ensuring the quality and usefulness of data. In the wild, this is often much easier said than done. Real world properties of datasets that you'll encounter contribute towards this being a time-consuming process. Data practitioners frequently must contend with the challenges of heterogeneous data formats and schemas, missing values, and poorly documented data sources, problems which are compounded as the size of data grows. It is normal for practitioners like data scientists and data analysts to spend more of their time in the process of data preparation, compared to the actual analysis of the data. It is a common complaint of data practitioners that their tooling gets in the way of effectively performing this work. DuckDB's versatile features and impressive performance helps improve the effectiveness of a wide-range of data-wrangling tasks, such as:

- Loading and combining data from different sources
- Mapping data from one raw data form into a conformed and often enriched format
- Summarizing datasets
- Handling missing or incomplete data

Data wrangling typically follows a set of general steps, beginning with extracting the data in a raw form from the data source, performing any cleaning required, such as removing duplicates and handling missing values, and then performing any transformations required, such as aggregating, sorting, or joining, in order to match a desired target structure, and then finally depositing the curated data into a target location for subsequent use. DuckDB can help us explore, transform, and validate raw datasets and is a flexible tool for data analysts to take messy data and create curated data ready for producing valuable insights and guiding business decisions.

Let's jump into using DuckDB for some data wrangling by analyzing user behavior from web server logs.

Imagine we are tasked with understanding the behavior of users who are interacting with a website. Our web server might be logging user requests by capturing each user interaction as a new line in a text file with a format established by our web server.

To begin our data wrangling, we need to understand our raw data source from our web server. Many web servers log user activity by creating a new record for each user interaction, such as browsing a new page, clicking a link, or uploading a file.

A typical record created by our web server has a log entry that looks like this:

```
2.182.248.122 user alice [26/Jan/2023:20:28:38 +0330] "GET blog.png"
200 321 "https://www.example.com" "Mozilla/5.0" "-" "en"
```

Note the spaces separating the fields and the use of the hyphen character (-) to designate missing entries. We can examine the fields that make up this line of data, which represents a user visiting our website:

- 2.182.248.122 is the **Internet Protocol** (**IP**) address of the client (remote host) that made the request to the server.
- user is the client identifier.
- alice is the identifier of the person using the web service.
- 26/Jan/2023:20:28:38 +0330 is the time of the request.
- GET blog.png is the request line from the client.
- 200 is the HTTP response status code returned to the client.
- 321 is the number of bytes returned to the client.
- en is the language code – in this case, it is the ISO 639-1 code for English.

Let's now look at how we can use DuckDB to help us wrangle this raw data and transform it into a more usable format for analyzing the behavior of the users interacting with our website.

Data import and manipulation

Our first data-wrangling task is to load the web server activity log into DuckDB. We begin by creating an empty web_log_text table to store the raw text of the web server log:

```
CREATE OR REPLACE TABLE web_log_text (raw_text VARCHAR);
```

This will create an empty table with a single column to store the raw lines of the web server. The CREATE OR REPLACE form of the CREATE TABLE statement instructs DuckDB to create a new table, overwriting any table that may exist with the same name.

To load our web server access.log file into the web_log_text table we can use the COPY statement.

```
COPY web_log_text FROM 'access.log' (DELIM '');
```

This might be a little counter-intuitive at first, as we are using DuckDB's CSV reader, which we saw in *Chapter 2*. While this isn't strictly a CSV file, it is a text file with each record occurring on a new line. We can leverage the CSV reader by treating it as though it has no field delimiters, which will result in each line of text being loaded as a distinct row. We can achieve this by providing the empty string for the value of the DELIM parameter. On running this statement, we will have loaded every line of the web server activity logs into rows of the web_log_text table. Let's peek at 10 rows of the table's contents:

```
SELECT *
FROM web_log_text
LIMIT 10;
```

This preceding query will give us a glimpse of the mess that we still have in each record from our raw log file:

```
151.239.2.90 - - [26/Jan/2023:20:28:37 +0330] "GET /image/18921/
productModel/50x50 HTTP/1.1" ...


69.162.124.229 - - [23/Jan/2023:11:51:13 +0330] "HEAD / HTTP/1.1" ...


151.239.2.90 - - [26/Jan/2023:20:28:37 +0330] "GET /image/62634/
productModel/50x50 HTTP/1.1" ...
```

This is some dense text to understand, as we have not yet extracted any structure from each record of our logs. The next bit of data wrangling will involve taming this unstructured text by extracting key fields through pattern matching. To do this, we'll need to dive into regular expressions.

Regular expressions

Regular expressions are sequences of characters that are used to match targeted patterns within text data programmatically. They are a powerful and concise way to work with text data, including tasks such as extracting, finding, replacing, or validating expected text values. Regular expression capabilities are found in many programming languages and data processing tools, including DuckDB.

Wikipedia has a good overview of regular expressions (https://en.wikipedia.org/wiki/Regular_expression) if you haven't encountered them before. We'll be using regular expressions as part of our data wrangling to extract specific values from our unstructured web log records so that we can create a structured table containing meaningful fields.

A regular expression consists of literal values (such as letters, digits, or punctuation symbols) and special characters and sequences that have specific meanings within the regular expression language that allow us to control the matching logic. For example, the \d special character matches a single digit, and [a-z] is a special sequence that matches any lowercase letter. The + special character indicates that the preceding expression will be matched one or more times, while the * special character indicates that the preceding expression will be matched any number of times, including zero matches. Let's put these together: the \d+ regular expression matches one or more digits, and the [a-z]* regular expression matches any lowercase letter, zero or more times.

We'll be using a handful of regular expressions to extract fields from our web log records. Let's take a look at one of them in detail, which will allow us to extract the IP address of the visiting user for each record, such as the 151.239.2.90 IP address we saw in the preceding section.

The ^[0-9\.]* regular expression will match a text sequence that occurs at the start of a string and contains zero or more repetitions of either a digit or a . character. The ^ symbol anchors our regular expression to match at the start of a string, and the [0-9\.] component matches any character that is either a digit from zero to nine or a dot (period), and then the following * symbol causes the preceding component to match zero or more times. This will easily match valid IP addresses such as 151.239.2.90 and 192.168.1.1. You may have noticed that this regular expression is a little overly permissive, also matching strings such as 123 and .45, which are not valid IP addresses. While we could refine our regular expression, this won't be an issue for our dataset, so we'll leave it as is.

The regexp_extract function

DuckDB provides the regexp_extract function, which extracts a matching substring from within supplied text according to a target regular expression pattern. We can use it in our data wrangling pipeline to extract the IP address from our web server records, as shown in the following query:

```
SELECT
    regexp_extract(raw_text, '^[0-9\.]*' ) AS client_ip
FROM web_log_text
LIMIT 3;
```

We can see that this query successfully extracts the IP address from each record in the web_log_text table:

```
151.239.2.90
69.162.124.229
151.239.2.90
```

Next, we'll see how we can extend this approach to extract other useful values and populate a table with these as fields.

Using regular expressions for text extraction

We're now ready to create a single query that selects from the `raw_text` table and uses four targeted regular expressions with the `regexp_extract` function to extract values for each record corresponding to the following fields:

- The visiting client's IP address
- The timestamp of the record
- The HTTP request method
- The language provided by the client

Here's our query, which we've limited to showing only five results:

```
SELECT regexp_extract(raw_text, '^[0-9\.]*' ) AS client_ip,
    regexp_extract(raw_text, '\[(.*)\]', 1) AS date_text,
    regexp_extract(raw_text, '"([A-Z]*) ', 1) AS http_method,
    regexp_extract(raw_text, '([a-zA-Z\-]*)"$', 1) AS lang
FROM web_log_text
LIMIT 5;
```

The `regexp_extract` function extracts the string that matches the given regular expression. We can place parentheses around the desired part of the regular expression we want to capture, which is referred to as a **capturing group**. By default, `regexp_extract` will return the whole segment of text matching the given regular expression. In our case, where we've used a matching group, we only want to extract the text from inside, so we pass the optional third argument to the `regexp_extract` function, to indicate which of the consecutive matching groups we want to return. We have passed 1 as this argument to indicate that we wish to extract the text from the first (and only) matching group in our regular expression. We have done this in some queries so that we only extract the desired subset of the matching text.

Running this query results in four columns being extracted for each record of our web activity log.

client_ip varchar	date_text varchar	http_method varchar	lang varchar
151.239.2.90	26/Jan/2023:20:28:37 +0330	GET	en
69.162.124.229	23/Jan/2023:11:51:13 +0330	HEAD	en
151.239.2.90	26/Jan/2023:20:28:37 +0330	GET	ja-JP
2.182.248.122	18/Jan/2023:20:28:38 +0330	POST	null
2.182.248.122	18/Jan/2023:20:28:38 +0330	POST	null

Figure 3.1 – Web server query results

Let's go through each of these new regular expression strings we're using to capture these new fields:

- `date_text`: `'\[(.*)\]'` matches any text that is enclosed in square brackets. The parentheses create a capturing group that defines the target text we want to extract. The dot symbol (`.`) matches any character except line breaks, and the asterisk (or star) character (`*`) means zero or more repetitions of the previous character. So, the regular expression can find dates such as `[26/Jan/2023:20:28:37 +0330]`, however only the portion inside the parenthesis will be returned, meaning the square brackets are not included in our extracted text.

- `http_method`: `'"([A-Z]*) '` matches a double quote followed by zero or more uppercase letters followed by a space. It uses a capturing group to extract the uppercase letters in a group. For example, it matches the `'GET'` string, extracting the `'GET'` substring.

- `lang`: `'([a-zA-Z\-]*)"$'` matches any sequence of alphabetic characters or hyphens at the end of a string, with the parentheses defining a capturing group that extracts language identifier strings like `'en'` and `'ja-JP'`.

While regular expressions can initially appear cryptic, they are worth becoming familiar with as they are a powerful way to search for, extract, or transform text data.

Now that we've established how to extract these fields from our web log data, let's persist these results into a table. We'll build a single query to execute each of these text-extraction regular expressions and create a corresponding `web_log_split` table like this:

```
CREATE OR REPLACE TABLE web_log_split AS
SELECT
    regexp_extract(raw_text, '^[0-9\.]*') AS client_ip,
    regexp_extract(raw_text, '\[(.*)\]', 1) AS http_date_text,
    regexp_extract(raw_text, '"([A-Z]*) ', 1) AS http_method,
    regexp_extract(raw_text, '([a-zA-Z\-]*)"$', 1) AS http_lang
FROM web_log_text;
```

This has created the `web_log_split` table, with the four columns we wanted to populate: `client_ip`, `http_date_text`, `http_method`, and `http_lang`.

We've made a good start on our data preparation journey; however, our data is not yet ready for analysis. In the next section, we'll cover the next set of data preparation steps, while also exploring different ways to build our data-wrangling pipeline with DuckDB.

Altering tables and creating views

Up to now, our data wrangling process has involved loading our web log data into a DuckDB table that stores log records as unstructured lines, followed by a reshaping step that builds a new structured table, with useful fields extracted as columns. In this section, we will cover the remaining two steps required to prepare the dataset for the analysis we want to perform: data type conversion and data enrichment. The first of these will ensure that each column has the appropriate data type for its contents, and the second will ensure that our columns are readily interpretable.

As we cover the remaining data preparation steps, we'll also look at different strategies for persisting these transformations. The first way involves applying stateful changes to an existing table. The second way involves using virtual tables, called **views**, to wrap up multiple transformations into a single convenient abstraction.

Converting strings to dates

Let's query the web_log_split table to see a sample of data and their corresponding data types:

```
SELECT *
FROM web_log_split
LIMIT 5;
```

We get the following result:

client_ip varchar	http_date_text varchar	http_method varchar	http_lang varchar
151.239.2.90	26/Jan/2023:20:28:37 +0330	GET	en
69.162.124.229	23/Jan/2023:11:51:13 +0330	HEAD	en
151.239.2.90	26/Jan/2023:20:28:37 +0330	GET	ja-JP
2.182.248.122	18/Jan/2023:20:28:38 +0330	POST	null
2.182.248.122	18/Jan/2023:20:28:38 +0330	POST	null

Figure 3.2 – A sample of records from our web_log_split table

We can see that we now have four VARCHAR columns. This, however, is an inappropriate data type for the http_date_text column, which should be a TIMESTAMP. Getting this data type right will allow us to perform user-behavior-oriented processing and analysis by time components, such as growth by month and highest activity by day.

We can use the DuckDB `strptime` function to parse a string and convert it into a timestamp value. It takes two arguments: a string representing the date and time and a format string that specifies how the target string is formatted. Using this function, we can convert text values such as `'26/Jan/2023:20:28:37 +0330'` into timestamps. The format string that we'll use to convert our timestamp strings is: `'%d/%b/%Y:%H:%M:%S %z'`. Note the final `%z` component; this corresponds to the substring `'+0330'`, which is the timezone offset, indicating that this timestamp is offset by three hours and thirty minutes from **Coordinated Universal Time** (**UTC**). As we'll discuss further in *Chapter 4*, DuckDB's `TIMESTAMP` data type does not store any timezone data. Because we are including timezone information in our `strptime` call, it will instead return a `TIMESTAMP WITH TIME ZONE` value, allowing us to retain this information.

```
SELECT client_ip,
    strptime(
        http_date_text,
        '%d/%b/%Y:%H:%M:%S %z'
    ) AS http_date,
    http_method,
    http_lang,
FROM web_log_split;
```

In this query, we've used the `strptime` function to parse the text in the `http_date_text` column and load them as `TIMESTAMP WITH TIME ZONE` values.

We'll need to update our data preparation process to take into account this data quality improvement. One way we could tackle this is to go back and change our table creation statement to include the `strptime` conversation logic. Generally speaking, this strategy is the preferred approach, as it will help keep your data cleaning process streamlined and reproducible. We'll take this opportunity, however, to illustrate how you can use DuckDB to update existing tables. We'll use this approach to update the `http_date` column in our existing `web_log_split` table.

Adding a new column

We will now alter the definition of the `web_log_split` table to add an additional column to store the activity date as `TIMESTAMP WITH TIME ZONE` values. We can add our new `http_date` column with this command:

```
ALTER TABLE web_log_split ADD COLUMN http_date
    TIMESTAMP WITH TIME ZONE;
```

This will create an additional `http_date` column in the `web_log_split` table with a data type of `TIMESTAMP WITH TIME ZONE`. We'll then use our `strptime` call we used earlier to populate the values of our new column, through the use of the `UPDATE` statement, which allows us to change column values in an existing table.

```
UPDATE web_log_split
SET http_date = strptime(
    http_date_text,
    '%d/%b/%Y:%H:%M:%S %z'
);
```

Let's query the table again to confirm that our new data preparation step has been applied:

```
SELECT client_ip,
    http_date,
    http_method,
    http_lang
FROM web_log_split;
```

We get the following results:

client_ip varchar	http_date timestamp with time zone	http_method varchar	http_lang varchar
151.239.2.90 69.162.124.229 151.239.2.90	2023-01-27 03:58:37+11 2023-01-23 19:21:13+11 2023-01-27 03:58:37+11	GET HEAD GET	en en ja-JP

Figure 3.3 – Web server results with the timestamp field

From this selection of our cleaned columns, we can see that the `http_date` column has the appropriate `TIMESTAMP WITH TIME ZONE` value. Thise correctly modeled timestamp will enable us to perform some analysis later on. Next, let's see what further data wrangling we can apply, this time to the `http_lang` column.

Creating a language lookup table

The http_lang column of our web_log_split table contains strings that indicate the language region of the user, but it is currently just a short language code with values such as en, ko, and fr. The next part of our data wrangling will be to enrich our dataset by converting these language codes into more descriptive language names such as English, Korean, and French. To do this, we'll create a language_iso language lookup table to map two-letter ISO 639-1 standard codes to more descriptive names and populate it using an appropriate mapping. Here is our table-creation step:

```
CREATE OR REPLACE TABLE language_iso (
    lang_iso VARCHAR PRIMARY KEY,
    language_name VARCHAR
);
```

Note that we've made the lang_iso column a PRIMARY KEY of this table, to ensure that there are no duplicate entries in our mapping of language codes to their more readable names. Now we'll insert the contents of the language_iso.csv file, which contains mappings of the two-letter ISO codes to more descriptive string values.

```
INSERT INTO language_iso
SELECT *
FROM read_csv('language_iso.csv');
```

This command has created and populated our language_iso lookup table that we can use to expand the short language codes.

Joining tables with DuckDB

We now have two tables, web_log_split, which contains web visit information, and the language_iso language lookup table. To map the language codes to their descriptive values, we need to combine these two tables via the language_iso field. We can do this using an **SQL join operation**, which allows us to combine data from two or more tables based on a related column between them.

DuckDB has comprehensive support for SQL join operations, with a range of types that are useful across different contexts. We'll just be considering the more traditional SQL join operations here. In *Chapter 10*, we'll look at more advanced join operations that DuckDB offers. Here are the classic SQL join operations:

- INNER JOIN: This returns records that have matching values in both tables. This would be appropriate if we knew there was always a match between our web_log_split table with the language_iso language lookup table.

- LEFT OUTER JOIN: This returns all records from the left table (regardless of matches) and the matched records from the right table. This type of join will be helpful if our web_log_split table has language codes not available in our language lookup table. The OUTER clause is optional but is usually included to help with readability.

- RIGHT OUTER JOIN: This returns all records from the right table and the matched records from the left table. This is logically equivalent to LEFT join – with the table names switched.

- FULL OUTER JOIN: This returns all records when there is a match in either the left or right table.

If SQL joins are new for you, there are some wonderful guides to help explain the styles of connecting tables. A good visual guide is available at https://www.atlassian.com/data/sql/sql-join-types-explained-visually.

For joining the language column to the web log table in order to use our mapping table, we're going to join our web_log_split table with the language_iso language lookup table. We'll use a LEFT OUTER JOIN ensuring that we join on the corresponding language code column in each table:

```
SELECT wls.http_date, wls.http_lang, lang.language_name
FROM web_log_split AS wls
LEFT OUTER JOIN language_iso AS lang
    ON (wls.http_lang = lang.lang_iso);
```

In this query, we are using a table alias for the tables; (web_log_split is aliased as wls and language_iso as lang). Using table aliases can be used as a substitute in place of referring to the table name when specifying columns. The LEFT JOIN command returns all the rows from the left table (web_log_split) and the matching rows from the right table (language_iso). As we've used an OUTER join, if there is no match, the record will be returned, but the columns from the right-hand-side table will have NULL values, which indicates empty fields. Using table aliases allows us to keep our query a little more concise, which is especially convenient in situations where we need multiple conditions in the WHERE clause.

After running this query, we get the query results as follows:

http_date timestamp with time zone	http_lang varchar	language_name varchar
2023-01-27 03:58:37+11	ja-JP	Japanese (Japan)
2023-01-20 03:58:39+11	fr	French
2023-01-20 03:58:39+11	es	Spanish
2023-01-19 03:58:38+11	**null**	
2023-01-27 03:58:53+11	ko	Korean
2023-01-27 03:58:53+11	en	English

. . .

As we can see, the `LEFT OUTER JOIN` has returned all the rows from the `web_log_split` table, returning a blank row when there is no matching language name for that code.

We have now established the logic needed for extracting dates as an appropriate data type and for expanding language codes to more descriptive values, however, you might have noticed that we have yet to persist data that combines both these enhancements into a table. In the next section, we'll look at an alternative to writing these results to a table, which will allow us to wrap up our processing logic into a single query that's convenient for data-wrangling workflows.

Creating a view

Our data processing steps have started to grow in complexity; we have multiple steps, which include updating an existing table, and we have yet to combine the last two steps. Instead of doing this by creating yet another table, it would be convenient if we could combine our two data enrichments into a single step that does not require updating an existing table, and without creating a new table, which we may not ultimately need.

This is where an SQL view, which DuckDB has support for via the `CREATE VIEW` command, can come in handy. It will enable us to wrap up our processing logic into a single step, without persisting an intermediate results table that would clutter up our database.

You can think of a view as a kind of virtual table that behaves much like a regular table in terms of how it can be queried, however, it does not store any data itself; instead, it will evaluate the results of its associated query on demand. The query behind a view can perform custom logic as needed, such as filtering, and joining across multiple tables, as we saw in the case of our language field expansion query.

Let's see how we can wrap up our previous two data processing steps into a single view that will streamline the subsequent use of this prepared data.

We'll use the `CREATE OR REPLACE VIEW` command to create a `web_log_view` view, which is composed of a query that includes our previous two processing steps of date conversion and language expansion via the mapping table:

```
CREATE OR REPLACE VIEW web_log_view AS
SELECT wls.client_ip,
    strptime(
        wls.http_date_text,
        '%d/%b/%Y:%H:%M:%S %z'
    ) AS http_date,
    wls.http_method,
    wls.http_lang,
    lang.language_name
FROM web_log_split AS wls
LEFT OUTER JOIN language_iso lang
    ON (wls.http_lang = lang.lang_iso);
```

Our view is now created and is associated with the query logic required to build our target results, but as we have not yet queried the view, DuckDB has not actually executed the underlying processing steps.

Let's use the DESCRIBE statement to inspect our view:

```
DESCRIBE web_log_view;
```

We get the following output:

column_name varchar	column_type varchar	null varchar	key varchar	default varchar	extra varchar
client_ip http_date http_method http_lang language_name	VARCHAR TIMESTAMP WITH TIME ZONE VARCHAR VARCHAR VARCHAR	YES YES YES YES YES			

Figure 3.4 – Description of the columns within the newly created view

We can see that even though the query hasn't been executed, DuckDB has registered the view with its column names and associated data types.

We can now use this view just as if it were a table, such as by querying it or using the COPY command to save it to a CSV, JSON, or Parquet file. Let's query this view:

```
SELECT *
FROM web_log_view
LIMIT 5;
```

We get the following results:

client_ip varchar	http_date timestamp with time zone	http_method varchar	http_lang varchar	language_name varchar
151.239.2.90	2023-01-27 03:58:37+11	GET	ja-JP	Japanese (Japan)
86.55.30.2	2023-01-20 03:58:39+11	GET	fr	French
5.121.166.61	2023-01-20 03:58:39+11	GET	es	Spanish
5.121.166.61	2023-01-20 03:58:39+11	GET	es	Spanish
5.121.166.61	2023-01-20 03:58:39+11	GET	es	Spanish

Figure 3.5 – Web traffic with the source language name

By querying web_log_view, we are in fact also causing DuckDB to execute the SQL query that defines the view, which, in this case, performs the string-to-date conversion and the language code expansion via the join.

> **Performance considerations of views**
>
> It's important to note that since views are not persistent, each time the view is queried, it will be evaluated again. This can have performance implications when the view contains complex logic or when the underlying dataset is large. Depending on the needs of your workflows, and the performance of your query, there are times when it may be more appropriate to persist the results of a query as a table instead of creating a view.

This illustrates how views can be used to encapsulate complex data processing steps in a convenient abstraction that does not require an intermediate table to be created. It's worth weighing up the tradeoff here – there will be an increase in the computing time each time a view is used, yet there is a reduction in the required storage space as you will not be persisting an intermediate result. Views provide an elegant mechanism for wrapping SQL that can still be flexibly consumed in a variety of ways.

Scripting SQL steps

We have been interacting with DuckDB by manually running commands one after another. This can understandably get a bit tiring (and error prone) if we want to run these steps again in the future.

In this chapter, we have created a raw table (web_log_text), loaded it from an audit file, split it into fields using a series of regular expressions (web_log_split), created a language lookup table and populated it (language_iso), and finally created a view to connect it together (web_log_view). That is a lot of steps! Fortunately, we can put multiple SQL commands in a text file and instruct DuckDB to process the steps sequentially with the .read dot command.

Have a look at the web_log_script.sql file. All of the commands you have used in this chapter are included in this script file. Before we run the script, we can get rid of existing tables and views by dropping them:

```
DROP VIEW IF EXISTS web_log_view;
DROP TABLE IF EXISTS language_iso;
DROP TABLE IF EXISTS web_log_split;
DROP TABLE IF EXISTS web_log_text;
```

We can recreate all these tables and views by running the SQL within the script with the .read dot command:

```
.read "web_log_script.sql"
```

After a few seconds of processing, our three tables and views should be recreated. We can query the view to confirm everything has been recreated:

```
SELECT *
FROM web_log_view
LIMIT 5;
```

This should show us that the data has been reloaded, converted, and correctly joined with the language lookup table.

We've done a lot of data wrangling to transform our raw web server logs into a more usable format. We are now ready to take this transformed data and perform some data analysis.

Aggregate functions and common table expressions

We have used data wrangling techniques to take our raw web server activity data and load it into DuckDB, parse it into meaningful fields, transform it into correct data types, and enrich it with added metadata. With these data processing steps complete, we'll now look at a type of operation we need for a core data analysis technique: the summarizing of large datasets by generating individual summary statistics of different fields that help us understand the shape of the underlying data. A common example is finding the average of a numerical field, which has the effect of converting a many-valued column into a single numerical value. This type of operation is commonly referred to as an aggregate function.

To ensure that we're ready for this section, we'll recreate all the necessary tables and views by executing the SQL from the `web_log_script.sql` script:

```
.read "web_log_script.sql"
```

Aggregate functions

To process our web activity, we will want to analyze large amounts of data together and summarize the results. Aggregate functions are useful for generating summary statistics that help us better understand the distribution of different columns within a dataset, such as the average and median values of a numerical column, the highest and lowest values (the *range*) of a numerical column, and the number of distinct values in a column. They are also useful for transforming data into different groupings, for example, by converting a table that contains records with hourly summary values, into a weekly summary of values.

An SQL aggregate function is a function that performs a calculation over multiple rows and combines them into a single returned value. For example, the `sum` function adds up all the values in a column and returns the total, the `max` function returns the maximum value in a column, and the `count` function returns the number of rows in a table or column.

For our first analysis step, let's find out the earliest and latest logged dates, along with how many rows we have in our dataset. We'll use the `min`, `max`, and `count` functions respectively to extract this information:

```
SELECT min(http_date) AS date_earliest,
    max(http_date) AS date_latest,
    count(*) AS web_log_count
FROM web_log_view;
```

By running this query, we can now see the date of our first and last records, along with the total number of records:

date_earliest timestamp with time zone	date_latest timestamp with time zone	web_log_count int64
2023-01-19 03:58:38+11	2023-02-01 03:58:54+11	500

Figure 3.6 – Web traffic visitor counts along with date range

We can see there are 500 rows exposed by the web_log_view view, spanning 12 days.

Bucketing dates

During our analysis of this web activity data, we may want to look at the user behavior at a daily, monthly, or yearly level. For example, we may want to truncate the 2023-01-19 03:58:38 timestamp to a daily level so we can combine all user activity for the day of 2023-01-19 together.

The time_bucket function is a DuckDB function that allows us to truncate a timestamp to the beginning of the nearest specified interval. This function can be useful when we want to group timestamps by arbitrary intervals. The function returns a timestamp value that marks the start of the interval that contains the source value. We can collapse our records into daily buckets like this:

```
SELECT http_date,
    time_bucket(interval '1 day', http_date) AS day
FROM web_log_view;
```

http_date timestamp with time zone	day timestamp with time zone
2023-01-27 03:58:37+11	2023-01-26 11:00:00+11
2023-01-20 03:58:39+11	2023-01-19 11:00:00+11
2023-01-20 03:58:39+11	2023-01-19 11:00:00+11

Note that the dates have been truncated, discarding the hours, minutes, and seconds. We will use the time_bucket function in the next query to see user behavior at a daily level.

Common table expressions

A **common table expression** (CTE) is a temporary named result set that can be used in a subsequent SQL statement. You can think of a CTE as functioning like a view that is only able to be used by the query it's defined in. A CTE can simplify complex queries by breaking them into smaller reusable parts, which can be used to improve the readability and maintainability of our SQL code. As we will see shortly, one of the reasons to use a CTE is to avoid repeating the same subquery logic multiple times in a larger query.

Imagine we wish to understand our daily user behavior on our website. For example, we may want to create a report of the most common languages used by visitors each day to our website. We can create a CTE that uses the `time_bucket` function to truncate the `http_date` values into days, meaning that we remove the time component of each value. We can then perform our aggregations against this CTE as if it were a view or a table.

A CTE is defined using the `WITH` keyword, followed by an expression name, the `AS` keyword, and then a query definition. After creating the CTE, we'll then follow up with our main aggregation query:

```
WITH web_cte AS (
    SELECT client_ip,
        time_bucket(interval '1 day', http_date) AS day,
        language_name
    FROM web_log_view
)
SELECT day, language_name, count(*) AS count
FROM web_cte
GROUP BY day, language_name
ORDER BY day, count(*) DESC;
```

The preceding snippet is a single SQL query that's composed of our CTE, which we've called `web_cte`, followed by a `SELECT` statement that queries the CTE. In order to get our daily language summaries, we've used a `GROUP BY` clause to group rows from the same day, along with the `count` aggregate function that will count the number of records within each day grouping. The `GROUP BY` clause is a common ingredient when performing aggregations. It collapses rows together into groupings of distinct values, allowing us to apply aggregate operations over groups of rows that match specified criteria. In our case, we want to bundle together all are rows sharing the same `day` and `language_name` into a single row, and count the number in each group using the count aggregate function. Finally, we use an `ORDER BY DESC` clause to sort the results in each group in descending order, with the earliest date at the beginning and the latest date at the end.

It's worth noting here that we have been able to use the day field of the CTE multiple times in the aggregate query – in the GROUP BY and ORDER BY clauses – without having to duplicate the logic in the CTE. The truncated query results are as follows:

```
|            day            |  language_name  |  count  |
|---------------------------|-----------------|---------|
| 2023-01-19 11:00:00+11    | English         |      18 |
| 2023-01-19 11:00:00+11    | Spanish         |       4 |
| 2023-01-19 11:00:00+11    | French          |       1 |
. . .
| 2023-01-21 11:00:00+11    | French          |      14 |
| 2023-01-21 11:00:00+11    | English         |      10 |
| 2023-01-21 11:00:00+11    | Korean          |       1 |
...
```

After running the query, we can see English speakers are the most prominent website users on the 19th of January, whereas French speakers are the most frequent on the 21st of January.

Pivot tables

We have been analyzing our web server logs and have seen the languages used by web visitors by day as a long vertical list of rows. There may be times when it is more convenient to transform rows into columns if we wish to see multiple values together on a row, or to create cross-tabular reports that display aggregated data from a different perspective. The SQL PIVOT command is used to transform rows into columns. Not all databases include SQL dialect support for pivot, so it is a noteworthy inclusion for DuckDB to have pivot available for creating more readable and compact reports, as well as for aggregating and summarizing data by different categories.

Let us now use a PIVOT statement to show the number of visits per country, where each language is in a separate column on a dated row:

```
WITH web_cte AS (
    SELECT time_bucket(interval '1 day', http_date) AS day,
        language_name
    FROM web_log_view
)
PIVOT web_cte ON language_name USING count(*);
```

The results are as follows:

day timestamp with time zone	Chinese int64	English int64	French int64	Hindi (India) int64	Italian int64
2023-01-26 11:00:00+11	1	67	117	3	0
2023-01-19 11:00:00+11	0	18	1	0	0
2023-01-20 11:00:00+11	0	10	2	0	0
2023-01-21 11:00:00+11	0	10	14	0	0
2023-01-22 11:00:00+11	0	13	25	0	3
2023-01-23 11:00:00+11	0	10	20	0	0
2023-01-24 11:00:00+11	0	3	12	0	0
2023-01-25 11:00:00+11	0	8	11	0	0
2023-01-27 11:00:00+11	3	0	2	0	0

Figure 3.7 – Truncated query results from PIVOT

This SQL command uses a CTE and a `PIVOT` clause to create a summary table of web log data. The `PIVOT` clause transforms `web_cte` into a new table that has one row for each day and one column for each language name. The `ON language_name` clause designates which column to split into columns, and the `USING count(*)` aggregate function is used to describe the values. Together, we have produced a count of HTTP requests for each language on each day.

This `PIVOT` command is useful in analysis situations where it is convenient to summarize data from different categories or groups in a single view.

We have seen a variety of ways to process data from a single table; let's now move on to querying data from more than a single table.

Joining data from multiple tables

We have learned the basics of data wrangling with DuckDB. Now, let's move on to a more detailed example. We're going to combine datasets from multiple tables in real-world taxi passenger-trip data, in order to analyze passenger movement and tipping behavior.

New York taxi data

The NYC Taxi and Limousine Commission provides a collection of data that contains information about the trips taken by yellow and green taxis in New York City. The dataset includes variables such as pickup and drop-off locations, dates and times, passenger counts, trip distances, fares, tips, tolls, and payment types. The dataset is publicly available and is interesting for analyzing traffic patterns and evaluating tipping behavior.

The dataset is updated monthly and can be accessed through the New York City website (`https://www.nyc.gov/site/tlc/about/tlc-trip-record-data.page`). For this exercise, we will be using the data from January 2023 for yellow taxi trip records. The data dictionary describing the attributes of the yellow taxi trip records is available on the NYC Taxi and Limousine Commission website (`https://www.nyc.gov/assets/tlc/downloads/pdf/data_dictionary_trip_records_yellow.pdf`).

Reading remote files with the HTTPFS extension

DuckDB has a number of extensions available that provide additional capabilities. We will be looking at extensions in more detail in *Chapter 5*, but for now, we will take advantage of DuckDB's `httpfs` extension to directly read data files from the New York City website. This extension is included in the standard DuckDB builds and is installed automatically when required. With the `httpfs` extension all DuckDB's file readers can consume from files hosted on HTTP endpoints. In this case, the NYC taxi data is provided in Parquet format, so we'll want to use the `read_parquet` function, which we'll use to directly select from the Parquet file corresponding to the January 2023 yellow taxi records and then save into a table named `trips`, all in a single SQL query:

```
CREATE OR REPLACE TABLE trips AS
SELECT *
FROM read_parquet('https://d37ci6vzurychx.cloudfront.net/trip-data/
yellow_tripdata_2023-01.parquet');
```

This query will take a few seconds to run, as the 45 MB Parquet file needs to be downloaded and then processed. Once that's done, have a look at the first few records in the `trips` table:

```
SELECT tpep_pickup_datetime,
    trip_distance,
    fare_amount,
    tip_amount,
    PULocationID,
    DOLocationID
FROM trips
LIMIT 10;
```

This provides us with a glimpse of some of the salient columns in this dataset.

tpep_pickup_datetime timestamp	trip_distance double	fare_amount double	tip_amount double	PULocationID int64	DOLocationID int64
2023-01-01 00:32:10	0.97	9.3	0.0	161	141
2023-01-01 00:55:08	1.1	7.9	4.0	43	237
2023-01-01 00:25:04	2.51	14.9	15.0	48	238
2023-01-01 00:03:48	1.9	12.1	0.0	138	7
2023-01-01 00:10:29	1.43	11.4	3.28	107	79
2023-01-01 00:50:34	1.84	12.8	10.0	161	137
2023-01-01 00:09:22	1.66	12.1	3.42	239	143
2023-01-01 00:27:12	11.7	45.7	10.74	142	200
2023-01-01 00:21:44	2.95	17.7	5.68	164	236
2023-01-01 00:39:42	3.01	14.9	0.0	141	107

Figure 3.8 – The first ten taxi trip records

Let's go through each of these columns, looking at what they encode:

- `tpep_pickup_datetime`: The date and time when the taxi trip started.
- `trip_distance`: The trip distance, in miles.
- `fare_amount`: The taxi fare for the trip, in dollars.
- `tip_amount`: Tip amount if paid by credit card, excluding cash tips, in dollars.
- `PULocationID`: A numeric identifier representing the pickup location.
- `DOLocationID`: A numeric identifier representing the drop-off location.

These field values are mostly easy to understand, however, the pickup and drop-off location values are not particularly useful to us as internal identifiers that this dataset uses to signify locations. We will need to do a bit more work to map the location identifiers to something human readable, so we can, for example, readily understand that a record corresponding to the `PULocationID` value of `161` is from the *Midtown Center* zone, in the *Manhattan* borough.

Location lookup

Reference data to understand the location identifiers is also made available by the NYC Taxi and Limousine Commission. We can use DuckDB to create and populate our location lookup table. First, we need to create a lookup table for locations like this:

```
CREATE OR REPLACE TABLE locations (
    LocationID int PRIMARY KEY,
    Borough VARCHAR,
    Zone VARCHAR,
```

```
        service_zone VARCHAR
);
```

This creates our locations table with a primary key on `LocationID` to ensure uniqueness. This is a guarantee enforced by DuckDB that every `LocationID` is a unique value appearing only once in the `locations` table.

We can now download the `taxi+_zone_lookup.csv` file to map a zone ID with a zone name and an associated borough name and insert the data into the `locations` table with the `read_csv` DuckDB function:

```
INSERT INTO locations(LocationID, Borough, Zone, service_zone)
SELECT LocationID, Borough, Zone, service_zone
FROM read_csv('https://d37ci6vzurychx.cloudfront.net/misc/taxi+_zone_
lookup.csv');
```

We insert the rows into the lookup table by pulling the CSV file directly from a remotely hosted file. It's worth a quick look at the data within our table:

```
SELECT LocationID, Borough, Zone
FROM locations
LIMIT 5;
```

LocationID int32	Borough varchar	Zone varchar
1	EWR	Newark Airport
2	Queens	Jamaica Bay
3	Bronx	Allerton/Pelham Gardens
4	Manhattan	Alphabet City
5	Staten Island	Arden Heights

Our location map shows some common locations in and around New York. We can now join our trip data to our location mapping table and create a table with the human-readable zone name for the pickup and drop-off location. Note that we are using the locations lookup table twice in this query: once mapped for pickup areas (aliased as `l_pu`) and again to map the drop-off areas (aliased as `l_do`):

```
CREATE OR REPLACE TABLE trips_with_location AS
SELECT t.*,
    l_pu.zone AS pick_up_zone,
    l_do.zone AS drop_off_zone
FROM trips AS t
LEFT JOIN locations AS l_pu
```

```
        ON l_pu.LocationID = t.PULocationID
LEFT JOIN locations AS l_do
        ON l_do.LocationID = t.DOLocationID;
```

The `trips_with_location` table is created, and we can query a few sample rows like this:

```
SELECT tpep_pickup_datetime,
    pick_up_zone,
    drop_off_zone,
    trip_distance
FROM trips_with_location
LIMIT 5;
```

The query result gives us an initial glimpse of the enriched ride data we have in our table.

tpep_pickup_datetime timestamp	pick_up_zone varchar	drop_off_zone varchar	trip_distance double
2023-01-01 00:32:10	Midtown Center	Lenox Hill West	0.97
2023-01-01 00:03:48	LaGuardia Airport	Astoria	1.9
2023-01-01 00:50:34	Midtown Center	Kips Bay	1.84
2023-01-01 00:21:44	Midtown South	Upper East Side North	2.95
2023-01-01 00:39:42	Lenox Hill West	Gramercy	3.01

Figure 3.9 – Taxi trips with zone names

Let's now do something a bit more interesting to see how people are spending their money on taxi trips.

Taxi trip cost and tipping

Now that we've collected our taxi trip data, let's have a look at the typical cost of a trip along with the most expensive journeys made.

We will create a query to look at the minimum, maximum, and average fare and tip amount for each day in January. We also want to see how generous people are when taking a taxi – so, let's work out the average daily tip, too. The tip amount field is only populated when a credit card is used as payment – so, we need to ensure only credit card transactions are used for determining the average tip percentage (cash tips are not included in the data).

To build our DuckDB SQL query to analyze the data of our taxi trips, we can construct a query with these parts:

- The `time_bucket` function to group the trips by day, based on the `tpep_pickup_datetime` column, which is the date and time when the passenger entered the taxi.

- A `CASE` expression inside the average function to calculate the average percentage of tip over fare amount for each day group, but only for those trips where the payment type was a credit card. The `payment_type` column is a numeric code that indicates how the passenger paid for the trip: 1 means credit card and 2 means cash. The average function ignores the null values and computes the average of the non-null ratios. The result is multiplied by 100 to convert it to a percentage.

- The `GROUP BY 1` clause specifies that all the aggregation functions (`count`, `min`, `max`, and `avg`) should be applied to each `day_of` bucket separately, rather than to the whole table.

- The `ORDER BY 1` clause sorts the results in date ascending order by the `day_of` column so that the earliest date will appear first.

Putting it all together, we can create a query like this:

```
SELECT time_bucket(interval '1 day',
       tpep_pickup_datetime) AS day_of,
    count(*) AS num_trips,
    min(fare_amount) AS fare_min,
    max(fare_amount) AS fare_max,
    avg(fare_amount) AS fare_avg,
    avg(tip_amount) AS tip_avg,
    avg(
        CASE WHEN Payment_type = 1
            THEN tip_amount / fare_amount
        END
    ) * 100 AS cc_tip_avg_pct
FROM trips_with_location
WHERE tpep_pickup_datetime BETWEEN
    '2023-01-20 00:00:00' AND '2023-01-29 23:59:59'
    AND fare_amount > 0
GROUP BY 1
ORDER BY 1;
```

Executing this query will take a moment and then should return a daily summary like this:

day_of timestamp	num_trips int64	fare_min double	fare_max double	fare_avg double	tip_avg double	cc_tip_avg_pct double
2023-01-20 00:00:00	108531	0.01	495.1	18.23	3.38	26.0
2023-01-21 00:00:00	111017	0.01	518.2	17.28	3.13	25.0
2023-01-22 00:00:00	88863	0.01	650.0	19.26	3.47	25.0
2023-01-23 00:00:00	88992	0.01	550.0	18.52	3.42	26.0
2023-01-24 00:00:00	103004	0.01	1160.1	17.79	3.35	26.0
2023-01-25 00:00:00	108328	0.01	400.0	17.34	3.32	28.0
2023-01-26 00:00:00	113979	0.01	600.0	18.34	3.48	26.0
2023-01-27 00:00:00	110720	0.01	500.0	18.07	3.38	26.0
2023-01-28 00:00:00	110653	0.01	400.0	17.31	3.15	25.0
2023-01-29 00:00:00	87373	0.01	598.7	19.31	3.49	25.0

Figure 3.10 – Daily taxi trip fare summary

It looks like New York is a busy place, with over 100,000 taxi trips taken on most days, with fewer trips on the weekend. The fare varies a bit, but it seems like the average fare is below $20. We can also see that passengers paying by credit card on most days tip around 25% of the metered fare.

It's worth noting the maximum fares are quite substantial compared to the average. It'll be interesting to explore the details of trips that are the most expensive each day – with the highest daily trip costing the passenger over $500.

Finding the most expensive trip with window functions

DuckDB supports window functions (also known as *analytic functions*), an SQL technique for advanced data analysis. Window functions can, for example, compute moving averages, cumulative totals, or ranks within a dataset. A window function operates on a subset of rows, allowing us to aggregate or rank a set of rows. The set of rows is referred to as a *window*, which we define with the OVER clause. Window functions provide a flexible mechanism for processing subgroups of data. See the DuckDB documentation on Window Functions (https://duckdb.org/docs/sql/window_functions.html) for further details for this flexible aggregation technique. We will also see the performance advantages of using window functions such as, LEAD and LAG, in *Chapter 4*.

For our exercise, we will use a DuckDB window function to locate the row that matches the maximum fare amount for each day.

We will construct a DuckDB query with a CTE that contains the columns from the trips_with_location table, as well as a new column called max_day_fare_amount. This column stores the maximum fare amount for each day, calculated by using the max window function over a partitioning that groups together trips on the same day. We can select the rows where fare_amount is equal to `max_day_fare_amount`, which means that these are the trips with the highest fare amount for each day. Altogether, our query to select the trip that is the most expensive each day looks like this:

```
WITH cte AS (
    SELECT twl.*,
    max(fare_amount) OVER (
      PARTITION BY
        time_bucket(INTERVAL '1 day', tpep_pickup_datetime)
    ) AS max_day_fare_amount
    FROM trips_with_location AS twl
)
SELECT tpep_pickup_datetime, pick_up_zone, drop_off_zone, fare_amount
FROM cte
WHERE fare_amount = max_day_fare_amount
    AND tpep_pickup_datetime BETWEEN
        '2023-01-20 00:00:00' AND '2023-01-29 23:59:59'
ORDER BY tpep_pickup_datetime;
```

Running the query with our window function shows the details of the trips that have the highest daily cost.

tpep_pickup_datetime timestamp	pick_up_zone varchar	drop_off_zone varchar	fare_amount double
2023-01-20 15:39:16	JFK Airport	Outside of NYC	495.1
2023-01-21 14:44:42	JFK Airport	Outside of NYC	518.2
2023-01-22 23:24:55	Clinton East	Clinton East	650.0
2023-01-23 23:01:23	Murray Hill	Outside of NYC	550.0
2023-01-24 12:43:44	JFK Airport	Outside of NYC	1160.1
2023-01-25 10:23:25	JFK Airport	Outside of NYC	400.0
2023-01-26 10:28:15	LaGuardia Airport	LaGuardia Airport	600.0
2023-01-27 01:42:45	Outside of NYC	Outside of NYC	500.0
2023-01-27 12:07:36	Jamaica	Jamaica	500.0
2023-01-28 23:00:50	Flatiron	Flatiron	400.0
2023-01-29 14:46:13	Clinton West	Outside of NYC	598.7

Figure 3.11 – Most expensive daily taxi fares

The query results show the trips with the highest fare amount for each day, in chronological order. It is worth noting that two trips are returned for 27th January – as they both match the highest spend for the day of $500.

By analyzing the most expensive daily taxi trips in New York, it's clearly worth considering sharing a taxi with a friend if you're starting your journey from LaGuardia Airport.

Summary

In this chapter, we learned how to use DuckDB effectively for performing a range of data manipulation activities, focusing in particular on how we can wrangle and clean data in order to prepare it for analysis. We took raw logs from a web server and converted them into a structured format that allowed us to perform some simple analysis on the data, such as seeing which types of users visit our website. We also performed the data processing required to prepare thousands of rows of taxi-trip data, allowing us to see how passengers travel in and around New York.

We have now seen a variety of core data manipulation steps and you should have a feel for how to use aggregate and window functions to discover interesting insights within large datasets using DuckDB.

Now that we have seen some techniques and strategies for using DuckDB to manipulate and analyze data with DuckDB, in the next chapter, we will further explore DuckDB techniques and features, this time looking at how you can improve the performance of your SQL queries.

4

DuckDB Operations and Performance

In this chapter, we dive into DuckDB operations and performance. We will explore some of the features and techniques that can help you understand and improve the speed and efficiency of your DuckDB workloads. We will start by discussing how DuckDB supports indexes to speed up queries on large tables, followed by exploring file operations on Parquet data files. During these explorations, we will cover how to analyze queries by looking at their query plans, as well as how we can use run-time profiling to inspect the timing and memory usage of our workloads.

We will also dive into working with timestamps, looking at the data types and features that DuckDB offers for handling times and time zones. We conclude by looking at how window functions can be used to efficiently process time-series data.

The chapter's exploration of these topics is structured as follows:

- Exploring DuckDB indexes
- Optimizing file read performance of DuckDB
- Working with timestamps
- Window functions and timestamps

Technical requirements

You will find the code examples and data files for this chapter in the chapter_04 folder in the book's GitHub repository, which can be found at https://github.com/PacktPublishing/Getting-Started-with-DuckDB/tree/main/chapter_04.

Exploring DuckDB indexes

A database index is a data structure that is used to improve the speed of data retrieval operations. Like many databases, DuckDB uses indexes to facilitate quick access to specific data within a database table by avoiding the need to perform an exhaustive search of every record. There are, however, some significant differences in how you should approach using indexing with DuckDB. Unlike many database systems, DuckDB takes care of many of the indexing operations automatically for you. This means you can spend less time on administrative tasks, such as index design, and focus more on the query logic needed to reach your desired outcomes.

DuckDB automatically creates a min-max index against every column that is a built-in general-purpose data type. This is also known as a **block range index** (**BRIN**) (`https://en.wikipedia.org/wiki/Block_Range_Index`). These indexes work by sampling a range of column data (a block) and storing the minimum and maximum values for each block. When evaluating queries, DuckDB uses BRIN indexes to determine whether a block contains any rows that satisfy a given condition on the indexed column without needing to scan the entire block. This substantially reduces the volume of data that must be examined by DuckDB and thereby speeds up your queries. The stored data structures used by this style of indexing also offer significant space savings compared to traditional database indexes such as B-trees.

DuckDB also supports **adaptive radix tree** (**ART**) indexes (`https://duckdb.org/docs/sql/indexes.html`), which use a highly tuned and space-efficient in-memory search tree. These indexes greatly speed up range scans, prefix lookup style queries, and joins. They are best used for high-cardinality columns. High cardinality for a database column means that it has many unique values and little repetition. For example, a column that stores user IDs, usernames, or email addresses would have high cardinality since each value is expected to be distinct from others.

An ART index is created automatically by DuckDB when you apply a PRIMARY KEY or UNIQUE constraint to a column. You can also manually create an ART index by using the CREATE INDEX command.

However, creating an index does introduce space overheads, so we don't want to add indexes to every column. It's worth understanding the situations where it makes sense to add an index. As a general principle, it is only worthwhile to explicitly create indexes in DuckDB to speed up queries for range queries and joins to other tables. We will use this section to explore what happens when we create indexes and also analyze the performance impacts of index creation.

Downloading the book review dataset from Kaggle

To explore the impact of adding DuckDB indexes, we are going to use a dataset of book reviews available on Kaggle, a website and community for data scientists and researchers.

To download the necessary dataset for this project, please follow these instructions:

1. Go to `https://www.kaggle.com/datasets/mohamedbakhet/amazon-books-reviews`.

2. Click on the **Download** button.

3. Kaggle will prompt you to sign in or to register. If you do not have a Kaggle account, you can register for one.

4. Upon signing in, the download will start automatically.

5. After the download is complete, unzip the `archive` zip into the `chapter_04` directory.

Once downloaded and unzipped, you should have two data files called `Books_rating.csv` and `books_data.csv`.

Preparing the book review dataset

The book reviews file `Books_rating.csv` has around 3 million book review records, but we want even more data for the remainder of this chapter. We are going to artificially expand this dataset by a factor of two by duplicating the data.

Our first activity will be to create a database sequence to uniquely identify the records. A **database sequence** is a type of object that generates a series of numeric values in a specified range and order. Sequences are often used to create unique identifiers for rows in a table or to generate sequential numbers for invoices or order numbers.

```
CREATE OR REPLACE SEQUENCE book_reviews_seq;
```

Once created, we can use the `book_reviews_seq` sequence in our next data preparation step.

In the below query, we're using DuckDB to both read the source CSV file and copy it to a local file using a single COPY command that has a nested query:

```
COPY (
    SELECT nextval('book_reviews_seq') AS book_reviews_id,
        Id AS book_id,
        Title AS book_title,
        Price AS price,
        User_id AS user_id,
        region,
        to_timestamp("review/time") AS review_time,
        cast(datepart('year', review_time) AS VARCHAR) AS review_year,
        "review/summary" AS review_summary,
        "review/text" AS review_text,
        "review/score" AS review_score
```

```
    FROM read_csv('Books_rating.csv')
    CROSS JOIN (
        SELECT range,
            CASE WHEN range = 0 THEN 'JP' ELSE 'US'
            END AS region
        FROM range (0, 2))
  ) TO 'book_reviews.parquet';
```

There's a bit going on in this command, so let's review what we have just done. We've parsed the Books_rating.csv CSV file, adding a new column called book_reviews_id with a sequence of numbers. We also renamed some poorly named columns and formatted them as needed. An SQL **cross-join** (also known as a **Cartesian join**) is used to return the Cartesian product of two tables. This means that every row of book data will be joined with each row of the dummy table without a matching condition, effectively doubling our dataset.

We are fabricating an artificial region column with the region values JP or US. Once completed, we write the resulting data to a Parquet file named book_reviews.parquet. We now have an enormous data file; let's see the various ways we can query this data in DuckDB.

> **Out of Memory Error**
>
> Depending on your hardware setup, you may encounter an out-of-memory error in these exercises, such as Error: Out of Memory Error: failed to allocate data of size. This error indicates that DuckDB is unable to hold the result set within the available memory. If you do encounter an error during these steps, you can instruct DuckDB to offload to disk as required by establishing a temporary disk file: SET temp_directory = 'temp.tmp';.

Reviewing and indexing the book review dataset

After running this initial data preparation step, we now have the book_reviews.parquet file created on disk. We'll then use the records in this file to create and populate a book_reviews table:

```
CREATE OR REPLACE TABLE book_reviews AS
SELECT *
FROM read_parquet('book_reviews.parquet');
```

To determine if adding an index is likely to be worthwhile, it can be helpful to review the distribution of data within the `book_reviews` table. We can use the `SUMMARIZE` command to display an approximate count of the number of unique values found in each column:

```
SUMMARIZE book_reviews;
```

We get the following result:

column_name varchar	column_type varchar	min varchar	max varchar	approx_unique varchar
book_reviews_id	BIGINT	1	6000000	5974834
book_id	VARCHAR	0001047604	B0064P287I	223629
book_title	VARCHAR	" Film technique, …	you can do anythin…	212573
price	DOUBLE	1.0	995.0	5996
user_id	VARCHAR	A00109803PZJ91RLT7…	AZZZZW74AAX75	1007952
region	VARCHAR	JP	US	2
review_time	TIMESTAMP	1969-12-31 23:59:59	2013-03-04 00:00:00	6222
review_year	VARCHAR	1969	2013	20
review_summary	VARCHAR	!	~~~~~~~~~~~~~~~~~~…	1548070
review_text	VARCHAR	The Tao of Muhamma	~~~~~~~~~~~~~~~~~~…	2002900
review_score	DOUBLE	1.0	5.0	5

Figure 4.1 – Results of the SUMMARIZE command

We can see that the `book_reviews_id` column has a little over 6 million unique values, making this a high-cardinality column and a good candidate for indexing. Conversely, the `review_year` column has only around 20 unique values, meaning indexing this column is unlikely to be beneficial.

In order to understand the nature of this dataset, we should go beyond just looking at summary statistics. We can retrieve and inspect a selection of records from the table using the DuckDB `SAMPLE` clause, which retrieves a randomly sampled subset of what would have been returned from a query. This can be useful for exploring a dataset faster, as it can be used to reduce the size of the dataset while still allowing for obtaining representative estimates of the data's characteristics and contents. It can also be used to manually inspect a small extract of the dataset. Here, we use the `SAMPLE` clause to randomly sample 10 records from our `book_reviews` table to give us an indication of the variety of data found in the book reviews within the Amazon Customer Reviews dataset:

```
SELECT *
FROM book_reviews
USING SAMPLE 10;
```

This gives us the following results:

book_reviews_id int64	book_id varchar	book_title varchar	price double	user_id varchar	region varchar
3060674	0767912926	Passing for Thin: …	11.21		GB
5958164	1561035025	The Scarlet Letter…	18.0	A1TB7TA59Q4W5M	US
2583758	081292987X	The Misunderstood …			GB
9457683	B000FC2QGE	Blink: The Power o…		A2O1FMNNGMOCM1	JP
8781953	B000PMCF1A	The Catcher in the…		A2Z55XB4086GRS	US
2921412	B000KTY7ZU	AN ACCEPTABLE TIME.			JP
7887850	0201615800	The Ultimate Windo…		A1E1F0WVTSRUAJ	US
11470634	0792717848	Like Water for Cho…		A2VX15DOYK27C0	JP
8517391	B000HE519E	Spiritual Interpre…		A7VIZK5EZ2JSI	JP
4958110	B000CQD40G	Kon-tiki		A3T7MC28NFTYCH	US

Figure 4.2 – Results from the sampling table

For the exploration of unfamiliar datasets, it's good practice to make use of sampling and not only rely on using a LIMIT clause to inspect portions of the data. One reason for this is that if the records of the dataset have a biased distribution, relying on LIMIT clauses to inspect slices of the data could give you a misleading sense of the nature of the whole dataset. In addition to this, the result sets returned by DuckDB (and SQL databases, more generally) do not have an inherent ordering. This means that when using a LIMIT clause, unless you are also specifying an ORDER BY clause, the order of the records returned up to the LIMIT quantity should be considered arbitrary, being influenced by contingent factors, such as the particular query plan that DuckDB generated for your query. Of course, this can still be useful, for example, when you need to visually inspect a small number of rows without concern for how representative they are of the entire dataset.

Explaining the query plan

To understand the potential performance improvements of using indexes, we need to introduce query plans. A database query plan is a representation of the steps that the database system will perform when executing a given query. A query plan consists of a tree of operators, such as scans, joins, aggregations, sorts, and projections, that process data from the underlying tables or views. We will use DuckDB's EXPLAIN statement to see the query plan and how it works internally.

Before we introduce any indexes to our book_reviews table, let's see the baseline behavior of having no manually created indexes. To get the query plan for a query, we simply prepend the keyword EXPLAIN. Let's look at the query plan that is used when selecting a specific user_id:

```
EXPLAIN
SELECT count(*)
FROM book_reviews
WHERE user_id = 'A1WQVN65FTJCJ6';
```

We get this result:

Figure 4.3 – Fragment of a query plan showing an index scan

The generated query plan is the verbose description of the processing steps. We can see the book_reviews table is being subjected to a scan of the entire table and the application of a filter, too, which requires user_id to match the value A1WQVN65FTJCJ6.

Let's also have a look at the query plan that is used when querying against a given year:

```
EXPLAIN
SELECT count(*)
FROM book_reviews
WHERE review_year = '2012';
```

We get the following result:

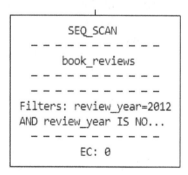

Figure 4.4 – Fragment of a query plan showing filter usage

Again, this shows us that the entire table is being scanned and the year filter is being applied. Let's now see how this behavior changes when we explicitly create indexes for this table.

Creating an index

Let's now create indexes for both the `user_id` and `review_year` columns of our `book_reviews` table to see what impact this has on retrieval performance for the two queries we ran in the previous section.

In DuckDB, we can manually create indexes on one or more columns of a table by using the `CREATE INDEX` statement:

```
CREATE INDEX book_reviews_idx_user_id ON book_reviews(user_id);
```

```
CREATE INDEX book_reviews_idx_year ON book_reviews(review_year);
```

We can query the `duckdb_indexes` metadata table to see the two created indexes:

```
SELECT *
FROM duckdb_indexes;
```

We get the following result:

database_name varchar	index_name varchar	table_name varchar	sql varchar
memory	book_reviews_idx_year	book_reviews	CREATE INDEX book_reviews_idx_year \nON book_reviews(year);
memory	book_reviews_idx_cu…	book_reviews	CREATE INDEX book_reviews_idx_customer_id \nON book_reviews(customer_id);

Figure 4.5 – List of indexes created

These two index creation commands have created non-unique indexes for the `user_id` and `review_year` columns, respectively. A unique index ensures that the values in the indexed column (or columns) are unique, whereas a non-unique index allows duplicate values in the indexed columns, which is what we want since users can write multiple reviews, and there will certainly be many records for each year.

Index performance

Let's now look at the performance impact of running queries against both these columns. Remember that we are anticipating that indexes will be most helpful for high-cardinality columns, such as `user_id`, which has a large number of distinct values, whereas we expect them to be less helpful for low-cardinality columns, such as "columns containing year values," which will have many duplicate values.

We'll start by rerunning the first of our queries to see the query plan when selecting a particular user_id now that this column has an index:

```
EXPLAIN
SELECT count(*)
FROM book_reviews
WHERE user_id = 'A1RRTLWXDOYER5';
```

We get the following result:

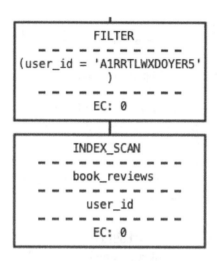

Figure 4.6 – A plan showing index usage

This query plan shows us that the index is indeed being used for the initial scan rather than a complete table scan. In this index scan, the query engine will find only the blocks that fall within the range of our manually created index for user_id. Only those rows then need to be processed by the next filtering step.

Now let's see what happens when we look at the query plan for the query against the review_year column now that it has an index:

```
EXPLAIN
SELECT count(*)
FROM book_reviews
WHERE review_year = 2012;
```

We get the following result:

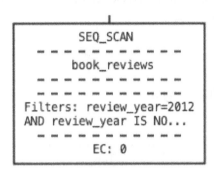

Figure 4.7 – A plan showing filter usage

In this case, the query plan is unchanged from the one that was produced when we ran our query against the non-indexed version of the table. This tells us that the query engine has chosen not to make use of the index for the review_year column for this query. This is due to the low cardinality of this column, where we have many duplicate year values occurring in this column. In this situation, the overheads of reading the index's tree data structure are likely to end up being less efficient than a simple sequential scan of the table, and thus, the query planner has chosen to use the default sequential table scan.

By comparing the query plans of a query against both a high-cardinality column and a low-cardinality column, both with and without corresponding column indexes, we can now see why it is not always beneficial to create indexes for columns. You may be wondering if there are any downsides of creating an index that is not used. This will be addressed in the next section, where we look at the storage implications of indexes.

Storage implications of indexes

It may seem tempting to always add indexes to our columns to speed up the retrieval of queries; however, there are trade-offs involved in adding indexes. Every index created will require additional storage for the data structures associated with the index. Additionally, these data structures will need to be updated with each insert, update, and delete operation, which can potentially have a negative impact on performance. We need to approach the decision to add any new index by weighing up the potential gains in query performance against the impact of additional storage and the performance penalty for index maintenance.

When focusing specifically on the storage impacts, we want to see specifically what happens to our table when we create an index on one of its columns. We previously saw how the EXPLAIN command produces a query plan that can be used to analyze how your query would be executed. However, in order to inspect diagnostics around the data storage being used by DuckDB, we need to use a different tool.

DuckDB provides a set of PRAGMA commands, which allow us to get and set internal database engine parameters, as well as view internal database diagnostics, such as the DuckDB version, the tables in the catalog, and disk and memory usage. **Pragmas** can be used to set execution behaviors in DuckDB, such as the display of a progress bar, the ordering behavior for NULLs, and the number of CPU threads to use. For a complete list of supported pragmas in DuckDB, see https://duckdb.org/docs/sql/pragmas.html. For our purposes here, we are interested in database size diagnostics that are provided by the database_size pragma.

In order to get a clean baseline, we first need to close our current in-memory database and connect to a new one. We can do this by running the following dot command in our DuckDB CLI:

```
.open
```

Let's start by getting the disk and memory size of the newly created and empty in-memory database:

```
PRAGMA database_size;
```

We get the following result:

database_name varchar	database_size varchar	block_size int64	total_blocks int64	used_blocks int64	free_blocks int64	wal_size varchar	memory_usage varchar	memory_limit varchar
memory	0 bytes	0	0	0	0	0 bytes	0 bytes	13.7GB

Figure 4.8 – Memory usage of a database

As you might have suspected, we can see that our freshly created database does not use disks or memory.

Now, we use the same command we used previously to create our book_reviews table:

```
CREATE OR REPLACE TABLE book_reviews AS
SELECT *
FROM read_parquet('reviews_original.parquet');
```

Now that we've created this table, let's run the pragma again to compare the storage usage after the creation of the table using DuckDB's in-memory database:

```
PRAGMA database_size;
```

We get the following result:

database_name varchar	database_size varchar	block_size int64	total_blocks int64	used_blocks int64	free_blocks int64	wal_size varchar	memory_usage varchar	memory_limit varchar
memory	0 bytes	0	0	0	0	0 bytes	10.1GB	13.7GB

Figure 4.9 – Memory usage without disk storage

In this instance, we have used 10.1 GB of memory to load the contents of the `reviews_original.parquet` file into the `book_reviews` table. This includes the automatically created BRIN indexes, which, as mentioned earlier, are created for every column.

Now, we'll create a **multi-column index** on the `book_reviews` table for the combination of the `region` and `review_score` columns. A multi-column database index is an index that is created for two (or more) columns of a table. It can potentially speed up queries for which the filter conditions involve all columns. We'll follow this by using another check of the current memory usage:

```
CREATE INDEX book_reviews_idx1 ON book_reviews(region, review_score);


PRAGMA database_size;
```

We get the following result:

database_name varchar	database_size varchar	block_size int64	total_blocks int64	used_blocks int64	free_blocks int64	wal_size varchar	memory_usage varchar	memory_limit varchar
memory	0 bytes	0	0	0	0	0 bytes	11.3GB	13.7GB

Figure 4.10 – Memory usage after index creation

The memory usage has increased to 11.3GB of memory. This suggests to us that approximately 1.2GB of memory was used to build this new index.

As we have seen, adding indexes consumes additional memory, so we should consider the downside of adding indexes needlessly in DuckDB. Is the potential performance improvement sufficient to justify a new index? The trade-off comes down to assessing if the additional memory used to create and maintain the index is worthwhile for the potential performance gained by improving retrieval speed via the additional index. Remember that DuckDB does a great job of using the automatically created BRIN indexes to maintain speedy data access with only modest storage overhead, so our advice is to be conservative when considering manually adding additional indexes. However, in the right circumstances, explicit column indexes can be an effective performance gain.

In the next section, we will continue our exploration of strategies for improving the performance of your DuckDB workloads; this time, we will look at the strategies you can use to speed up reading data from large collections of Parquet files.

Optimizing the file read performance of DuckDB

Consuming data from files stored on a disk is a common pattern in the data world, and we have seen many examples where DuckDB is used to read directly from file and network paths. It's worth understanding some of the ways we can maximize performance when reading datasets from files stored on a disk. We will be exploring the clever ways DuckDB can optimize the reading of large datasets stored in files and the techniques for arranging them on a disk to improve reading speeds.

File partitioning

We learned about Hive partitioning in *Chapter 2*, a technique that allows you to organize files on disk by dividing a single table into smaller logical tables based on the values of a particular column. This column is known as the **partition key**, which frequently takes the form of a date component, dividing up records into different time periods, such as months and years. In a similar way to how DuckDB's BRIN indexes leverage block min-max metadata to scan less data during query time, this partitioning can be used by DuckDB to dramatically reduce the amount of data that needs to be fetched and processed when reading from cloud object storage.

Let's now look more closely at the performance benefits of loading data from a partitioned directory structure. We will take our Amazon Customer Reviews dataset and save it to disk in Parquet files that are partitioned into directories by both `review_year` and `region`. To help us more clearly identify the effect of using indexes, we're going to deliberately prevent DuckDB from being able to parallelize its query execution over multiple threads. We'll limit the database to use a single thread for this repartitioning exercise by setting the `threads` configuration parameter using the `SET` statement:

```
SET threads TO 1;

COPY (
    SELECT *
    FROM read_parquet('book_reviews.parquet')
) TO 'book_reviews_hive' (
    FORMAT parquet,
    PARTITION_BY (review_year, region),
    OVERWRITE_OR_IGNORE true
);
```

This query loads the data from `book_reviews.parquet` and writes them to the disk as a collection of Parquet files in a directory named `book_reviews_hive`, which is partitioned into subdirectories based on the review year, with those directories, in turn, being partitioned into region subdirectories as follows:

```
|-- review_year=1996
|    |-- region=JP
|    |    -- data_0.parquet
|    -- region=US
|        -- data_0.parquet
.  .  .
|-- review_year=2015
|    |-- region=JP
|    |    -- data_0.parquet
|    -- region=US
|        -- data_0.parquet
```

As we can see from the directory structure of our now Hive-partitioned collection of Parquet files, the columns by which the dataset is partitioned have been encoded in the directory path of each file.

By writing a dataset to disk using this partitioning, we are now able to take advantage of another performance optimization available when consuming from Parquet files. When querying records in a Hive-partitioned dataset, DuckDB can use the filter predicate found in the query's WHERE clause to read only a subset of Parquet files for which the partition keys match the target filter. When dealing with large datasets, being able to exclude portions of the dataset can dramatically reduce query times by avoiding unnecessary file reads.

Let's run some tests to see this behavior with our recently partitioned data. Firstly, we can set a timer to show the elapsed time:

```
.timer on
```

Let's imagine we wanted to run a query that retrieves reviews from Japan that were submitted in 2012. When creating a query to read from our partitioned dataset, we can indicate to DuckDB that it should read the Parquet files as a Hive-partitioned dataset with the hive_partitioning parameter of the read_parquet function:

```
SELECT *
FROM read_parquet(
    'book_reviews_hive/*/*/*.parquet',
    hive_partitioning=true
)
WHERE review_year = '2012' AND region = 'JP';
```

Run Time (s): real 0.548

When executing this query, DuckDB is able to leverage the partitioning and, consequently, will scan only the files that satisfy the two conditions in the WHERE clause: review_year = 2015 AND region= 'JP'. Behind the scenes, DuckDB will use the predicate and rewrite the query to only read files matching the book_reviews_hive/review_year=2015/region=JP/*. parquet path.

Be sure to note how long the partition-aware query takes to execute on your computer, as we will compare that with the results of our next query. This time, we'll run the same query but against our non-partitioned reviews_original.parquet file:

```
SELECT *
FROM read_parquet('book_reviews.parquet')
WHERE review_year = '2012' AND region = 'JP';
```

Run Time (s): real 1.062

As we can see, this query took longer to run than its Hive-partitioned counterpart. The DuckDB query engine had to perform additional disk reads here, making it slower than was possible with Hive-partitioning.

It's worth considering adopting Hive partitioning in situations where you will be frequently querying a large dataset that is stored on disk and would benefit from lower latency queries. This makes it a particularly common pattern in data lakes, with Parquet or CSV files stored using Hive partitioning on cloud object storage, which can then be more readily queried by SQL query engines, such as Amazon Athena and Trino.

Parquet predicate pushdown

As we have already seen, DuckDB's built-in Parquet reader enables us to run queries directly against Parquet files without any import or analysis steps. When querying Parquet files like this, DuckDB is able to employ several optimizations that are made possible by the Parquet format, which enables DuckDB to increase the efficiency of queries. One we have already mentioned is that DuckDB will use Parquet metadata to read only the columns in the file that are required for a given query. In addition to this, Parquet columns are partitioned into blocks of row groups, and metadata is stored for each block, including min/max values and the number of NULL values. DuckDB is able to leverage this block metadata to skip reading blocks within a file that do not match the query criteria. **Predicate pushdown** evaluates the predicates (the WHERE conditions) at a lower level and discards non-matching Parquet files. That is, it is possible for DuckDB to selectively read blocks of data and avoid unnecessary read operations from Parquet files.

DuckDB also reads Parquet data as a stream, meaning that it can start operating on data before it has completed its read operations on a file. This, combined with DuckDB's ability to skip columns and rows that are not needed for a query, means that it is possible to query large Parquet files that do not fit in memory.

In the next exercise, we'll look at DuckDB's ability to efficiently query very large Parquet files.

Run-time profiling exercise

Earlier, we looked at using the EXPLAIN command to view a DuckDB query plan. This was useful for looking at the logical steps, but let's now get some further details by enabling run-time profiling. We can enable run-time profiling to see both how much time and how many rows (cardinality) each step of a statement processes. We can use run-time profiling to measure Parquet performance optimizations.

We are going to use a single thread and enable the writing of our run-time profiling to a file called profile_with_pushdown.log using the following commands:

```
SET threads TO 1;
PRAGMA enable_optimizer;
PRAGMA enable_profiling;
PRAGMA profiling_output = 'profile_with_pushdown.log';
```

Now run the following table creation, followed immediately by closing the profiling capture:

```
CREATE OR REPLACE TABLE book_reviews_1970_JP AS
SELECT region,
    review_summary,
    review_text,
    review_time,
    review_year
FROM read_parquet('book_reviews.parquet')
WHERE region = 'JP' AND review_year = '1970';
PRAGMA disable_profiling;
```

We have created a table and created a run-time profile of the activity steps. We can open the profile file `profile_with_pushdown.log` using our favorite text file viewer. At the bottom of this profiling file is a count of the number of rows read from the file by the Parquet reader, as shown in the following screenshot:

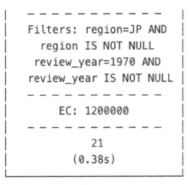

```
| - - - - - - - - - - |
|  Filters: region=JP AND  |
|    region IS NOT NULL    |
|   review_year=1970 AND   |
| review_year IS NOT NULL  |
|                          |
| - - - - - - - - - - |
|       EC: 1200000        |
| - - - - - - - - - - |
|           21             |
|         (0.38s)          |
```

Figure 4.11 – Profile output with elapsed time

In this example, the filters `region = 'JP' AND review_year = '1970'` were applied while reading the file, and only 21 rows were returned to DuckDB in 0.38 seconds.

Now, let's do the same activity, but this time, we want to disable DuckDB's ability to use the Parquet skip block optimization. We can disable this capability temporarily by executing the `disable_optimizer` pragma:

```
PRAGMA disable_optimizer;
PRAGMA enable_profiling;
PRAGMA profiling_output = 'profile_without_pushdown.log';

CREATE OR REPLACE TABLE book_reviews_1970_JP AS
SELECT region,
    review_summary,
```

```
      review_text,
      review_time,
      review_year
FROM read_parquet('book_reviews.parquet')
WHERE region = 'JP' AND review_year = '1970';


PRAGMA disable_profiling;
```

With all these commands running, the run-time profile will be written to `profile_without_pushdown.log`. At the bottom of this profiling file is a count of the number of rows read from the file by the Parquet reader:

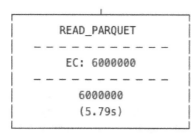

Figure 4.12 – Profile output with increased elapsed time

In this example, the entire file of just over 6 million rows was returned to DuckDB, a noticeably longer time of 5.79 seconds.

It is striking the optimizations that are available in DuckDB when processing Parquet data files. Run-time profiling is able to help us quantify the efficiency possible for data storage and retrieval.

Now we have explored file and data storage optimizations, let's turn our attention to effectively working with date and time in DuckDB.

Working with timestamps

Processing and storing temporal data are common activities in data analysis and data management. DuckDB provides data types and functions specifically for working with data that represent times, dates, intervals, and timestamps. Timestamps are combinations of dates and times that are frequently used in data analytics contexts, especially when working with time series datasets. Timestamps can sometimes be a misunderstood concept, so in this section, we'll introduce some principles for thinking about working with timestamp data in DuckDB before jumping into some examples of working with timestamps.

Timestamps represent distinct points in time, often referred to as **instants**. DuckDB has two data types for representing timestamps: TIMESTAMP and TIMESTAMP WITH TIME ZONE.

The latter type is also available as the alias TIMESTAMPZ, which we'll use from here on for brevity. These are both used to store date and time information. As you might have guessed, the difference between them lies in how they handle time zone information.

Under the hood, DuckDB uses the common practice of representing timestamps as the number of microseconds elapsed since 1970-01-01 00:00:00 UTC, a fixed moment in time that is commonly referred to as the **Unix epoch**, or just the **epoch**. A negative timestamp value indicates an instant that is earlier than the epoch, and a positive value indicates an instant that occurs after. This, of course, wouldn't be a particularly useful way to represent timestamps to a user; DuckDB renders timestamps in a human-friendly representation, such as 2023-06-08 11:36:21. While it may not feel like you're working with epoch-based timestamps, it is important to understand that this is how DuckDB represents timestamps internally.

TIMESTAMP

DuckDB's TIMESTAMP data type represents a specific point in time without consideration of time zone information. As mentioned, a TIMESTAMP value is stored as an instant: the number of microseconds since the epoch. In order to process and render this epoch value in a useful way, DuckDB chunks it up into years, months, days, hours, minutes, and seconds. To do this, DuckDB needs to know where this **binning** (chunking into years, months, hours, and so on) should be performed, relatively speaking. This needs to be done because the same instant in time can fall into different days, depending on where on the Earth you are. The TIMESTAMP data type does not, however, encode any time zone information about the instant, so it assumed that the timestamp should be binned according to the time zone configured by the database. To summarize, TIMESTAMP values do not encode any time zone information, and it is up to the application that is interacting with DuckDB to interpret and display it correctly based on the desired time zone.

Consider the moment the first time a human stepped on the Moon: *"Commander Neil Armstrong became the first person to step onto the Moon's surface on July 21 1969 at 02:56 UTC"* (https:// en.wikipedia.org/wiki/Apollo_11).

If your DuckDB session client is set to UTC and you insert the time a human first stepped on the Moon (1969-07-21 02:56:00) into a column using the TIMESTAMP data type, it will be stored as -14173440 seconds before the epoch.

```
SELECT TIMESTAMP '1969-07-21 02:56:00' AS moonstep;
```

```
|        moonstep        |
|        timestamp       |
|------------------------|
| 1969-07-21 02:56:00    |
```

When you retrieve the value, it will be returned as `1969-07-21 02:56:00`.

TIMESTAMPTZ – Timestamps with time zones

As with `TIMESTAMP`, the `TIMESTAMPTZ` type stores a specific point in time, but it also stores a time zone alongside this, which is used for binning epoch values. This DuckDB post on time zones (`https://duckdb.org/2022/01/06/time-zones.html`) has this to say about time zone data types:

"The SQL standard defines temporal data types qualified by WITH TIME ZONE. This terminology is confusing because it seems to imply that the time zone will be stored with the value, but what it really means is "bin this value using the session's TimeZone setting."

For example, consider the moment a human first stepped on the Moon from the perspective of a journalist writing an article for the New York Times. If we imagine that journalists used computers rather than typewriters back then, their system would be set to New York time, which is UTC-4. From their perspective, the timestamp will be the value `'1969-07-20 22:56:00-04'` when entered into a column using the `TIMESTAMPTZ` data type. The important thing to note here is that the timestamp is the same moment in time regardless of the time zone:

```
SELECT TIMESTAMPTZ '1969-07-20 22:56:00-04' AS moonstep_z;
```

When you retrieve the value, it will be returned with the time zone offset included, such as `'1969-07-20 22:56:00-04'`.

Exploring timestamps with time zone functions

Let's explore timestamps in DuckDB by representing the major events of the Apollo 11 American spaceflight that first landed humans on the Moon in July 1969.

Let's see how we can describe this significant event in history and, by doing so, see the effect of using different data types when describing the same event. We will start by creating a table using both `TIMESTAMP` and `TIMESTAMPTZ` data types and set our environment to the UTC time zone:

```
SET TimeZone = 'UTC';

CREATE OR REPLACE TABLE timestamp_demo (
    col_ts TIMESTAMP,
    col_tstz TIMESTAMPTZ
);
```

We will now insert the same moment of the first step on the Moon into each of the TIMESTAMP and TIMESTAMPTZ columns:

```
INSERT INTO timestamp_demo (col_ts, col_tstz)
    VALUES('1969-07-21 02:56:00', '1969-07-21 02:56:00');
```

We can retrieve these values and see the representation from the perspective of our current UTC time zone setting:

```
SELECT current_setting('timezone') AS tz,
    col_ts,
    extract(epoch FROM col_ts) AS epoc_ts,
    col_tstz,
    extract(epoch FROM col_tstz) AS epoc_tstz
FROM timestamp_demo;
```

We get the following result:

tz varchar	col_ts timestamp	epoc_ts int64	col_tstz timestamp with time zone	epoc_tstz int64
UTC	1969-07-21 02:56:00	-14159040	1969-07-21 02:56:00+00	-14159040

Figure 4.13 – Timestamps as offsets from the epoch

With the time zone set to UTC, we can see the timestamp field is returned as 1969-07-21 02:56:00, with the timestamp with time zone field showing the same value, with the current time zone offset included: 1969-07-21 02:56:00+00. It is worth noting that both values have an internal epoch offset of -14159040, which is a value of 163 days prior to the beginning of 1970.

Now, let's change the time zone to New York time and execute the same query:

```
SET TimeZone = 'America/New_York';
SELECT current_setting('timezone') AS tz,
    col_ts,
    extract(epoch FROM col_ts) AS epoc_ts,
    col_tstz,
    extract(epoch FROM col_tstz) AS epoc_tstz
FROM timestamp_demo;
```

We get the following result:

tz varchar	col_ts timestamp	epoc_ts int64	col_tstz timestamp with time zone	epoc_tstz int64
America/New_York	1969-07-21 02:56:00	-14159040	1969-07-20 22:56:00-04	-14173440

Figure 4.14 – Time, with time zone expressed against EPOC

With the time zone set to New_York, the timestamp is still shown as UTC 1969-07-21 02:56:00, which is unchanged after altering our local time zone. However, the timestamp from the column with a time zone type is displayed as the previous day 1969-07-20 22:56:00-04, including a time zone offset of -04. The impact of using the TIMESTAMPTZ data type is that DuckDB acknowledges that the current time zone setting should be used when rendering this type.

Binning and functions using time zone functions

The implications of using a TIMESTAMP and TIMESTAMPTZ data types do not just relate to how timestamps are visually rendered. For example, DuckDB will treat these data types differently in grouping operations such as binning. Take a situation where we wish to identify the day of the week when the Moon landing occurred:

```
SELECT current_setting('timezone') AS tz,
    col_ts,
    dayofmonth(col_ts) AS day_of_month_ts,
    dayname(col_ts) AS day_name_ts,
    col_tstz,
    dayofmonth(col_tstz) AS day_of_month_tstz,
    dayname(col_tstz) AS day_name_tstz
FROM timestamp_demo;
```

We get the following result:

tz varchar	col_ts timestamp	day_of_month_ts int64	day_name_ts varchar	col_tstz timestamp with time zone	day_of_month_tstz int64	day_name_tstz varchar
America/New_York	1969-07-21 02:56:00	21	Monday	1969-07-20 22:56:00-04	20	Sunday

Figure 4.15 – Binning time with time zones

With the time zone set to New_York, the timestamp still describes the Moon landing event as being in relation to UTC: Monday 21. The timestamp with the time zone would describe the same moment from the perspective of the current New York time zone and equates the moment as Sunday 20.

Time and window functions

We have seen how to represent a single moment in time using DuckDB's TIMESTAMP and TIMESTAMPTZ data types. Now, let's explore durations—the period of time that passes between timestamps—in an example as follows.

The Apollo 11 mission

Consider the chronology of events from the Apollo 11 mission, which was a historic achievement that marked the first time humans landed on the Moon. The mission was launched by NASA on July 16, 1969, and consisted of three spacecraft: the Command Module, the Service Module, and the Lunar Module. The crew of Apollo 11 included Neil Armstrong, Michael Collins, and Buzz Aldrin. The journey to the Moon took 4 days. On July 20, Armstrong and Aldrin separated the Lunar Module from the Command Module and descended to the lunar surface while Collins remained in orbit. At 20:17 UTC, the Lunar Module touched down on the Moon, and 6 hours later, Armstrong became the first human to step out of the Lunar Module and onto the Moon, followed by Aldrin. They spent about two and a half hours outside the Lunar Module. On July 21, Armstrong and Aldrin re-entered the Lunar Module and lifted off from the Moon at 17:54 UTC. They rendezvoused and docked with Collins in the Command Module. On July 24, the Command Module re-entered Earth's atmosphere and splashed down in the Pacific Ocean at 16:50 UTC.

Let's use the events of the Apollo 11 mission to understand who spent the most time on the Moon. We can load the activities from Neil Armstrong, Michael Collins, and Buzz Aldrin into the apollo_events table:

```
CREATE OR REPLACE TABLE apollo_events AS
SELECT *
FROM read_csv(
    'apollo.csv',
    timestampformat='%d/%b/%Y %H:%M',
    columns={
        'event_time': 'TIMESTAMP',
        'astronaut': 'VARCHAR',
        'event_description': 'VARCHAR',
        'astronaut_location': 'VARCHAR'
    }
);
```

With the table loaded, we can query the apollo_events table to review the activities of Neil Armstrong:

```
SELECT *
FROM apollo_events
WHERE astronaut = 'Neil Armstrong'
ORDER BY event_time;
```

We get the following result:

event_time timestamp	astronaut varchar	event_description varchar	astronaut_location varchar
1969-07-14 21:00:00	Neil Armstrong	Enterered Apollo 11 command module	Command module
1969-07-20 14:32:00	Neil Armstrong	Entered Apollo Lunar Module	Lunar module in space
1969-07-20 20:17:00	Neil Armstrong	Apollo Lunar Module lunar landing	Lunar module on moon
1969-07-21 02:56:00	Neil Armstrong	First step on lunar surface, That's one small step for a man	Moon
1969-07-21 05:09:00	Neil Armstrong	Returned inside Apollo Lunar Module	Lunar module on moon
1969-07-21 17:54:00	Neil Armstrong	Apollo Lunar Module lunar liftoff ignition	Lunar module in space
1969-07-21 22:52:00	Neil Armstrong	Entered Command module	Command module
1969-07-24 17:29:00	Neil Armstrong	Egress	Recovery ship

Figure 4.16 – Apollo events for Neil Armstrong

With the data loaded, we can move on to finding the duration of the Apollo 11 mission.

Interval data

The INTERVAL data type provides a way of representing a duration of time in DuckDB. If we subtract a timestamp from another timestamp, we obtain an interval describing the difference between the timestamps. For example, we can work out how long Neil Armstrong spent on the surface of the Moon by looking at the rows of data representing the moment he stepped onto the lunar surface and the moment in time when he returned inside the Lunar module. We can determine the interval by subtracting the former timestamp from the later timestamp using a subtraction operator (-):

```
SELECT TIMESTAMP '1969-07-21 05:09:00'
    - TIMESTAMP '1969-07-21 02:56:00' AS interval_on_moon;
```

This returns the duration of time on the Moon as an INTERVAL data type:

```
| interval_on_moon |
|     interval     |
|                  |
| 02:13:00         |
```

This shows that Neil spent 2 hours and 13 minutes on the lunar surface. Let's look at creating a query for the apollo_events table to see the duration of all the significant lunar activities.

The LEAD and LAG window functions

The major events of the Apollo 11 mission are stored as rows in the apollo_events table. To determine the duration of an activity, we need to find the difference between the timestamp value of the current row and the timestamp of the next row; that is, our query will be performing calculations over a group of rows and will calculate a result for the period of time between one row and the next. If there is no next row for an astronaut, it returns NULL.

The LEAD and LAG functions are window functions that allow you to access data from other rows in the same result set. We are going to use the LEAD function to return the timestamp value from a row that follows the current row by a specified offset. We can use the LEAD function to compare values across rows to calculate the interval between events. Similarly, the LAG function provides access to a row that comes before the current row.

We will create an SQL view with a query to retrieve information about the duration of events during the Apollo mission:

```
CREATE OR REPLACE VIEW apollo_activities AS
SELECT event_description,
    event_time,
    astronaut,
    astronaut_location,
    LEAD(event_time, 1) OVER (
        PARTITION BY astronaut
        ORDER BY event_time
    ) AS end_time,
    end_time - event_time AS event_duration
FROM apollo_events;
```

The LEAD function returns the value of a column from the next row in a partition. In this case, it returns the event time of the next event for each astronaut. The OVER clause defines how to partition and order the rows for the window function. In this case, it is partitioned by astronaut and is ordered by event time. This means that for each astronaut, the events are ordered chronologically, and the end_time column shows the start time of the next event. We evaluate end_time - event_time to calculate the duration of each Apollo event by subtracting the start time from the end time as an event_duration interval.

We can query the apollo_activities view to see the activities of Neil Armstrong:

```
SELECT *
FROM apollo_activities
WHERE astronaut = 'Neil Armstrong'
ORDER BY astronaut, event_time;
```

We get the following result:

event_description varchar	event_time timestamp	astronaut varchar	astronaut_location varchar	end_time timestamp	event_duration interval
Enterered Apollo 11 command module	1969-07-14 21:00:00	Neil Armstrong	Command module	1969-07-20 14:32:00	5 days 17:32:00
Entered Apollo Lunar Module	1969-07-20 14:32:00	Neil Armstrong	Lunar module in space	1969-07-20 20:17:00	05:45:00
Apollo Lunar Module lunar landing	1969-07-20 20:17:00	Neil Armstrong	Lunar module on moon	1969-07-21 02:56:00	06:39:00
First step on lunar surface, That's one small step for a man	1969-07-21 02:56:00	Neil Armstrong	Moon	1969-07-21 05:09:00	02:13:00
Returned inside Apollo Lunar Module	1969-07-21 05:09:00	Neil Armstrong	Lunar module on moon	1969-07-21 17:54:00	12:45:00
Apollo Lunar Module lunar liftoff ignition	1969-07-21 17:54:00	Neil Armstrong	Lunar module in space	1969-07-21 22:52:00	04:58:00
Entered Command module	1969-07-21 22:52:00	Neil Armstrong	Command module	1969-07-24 17:29:00	2 days 18:37:00
Egress	1969-07-24 17:29:00	Neil Armstrong	Recovery ship		

Figure 4.17 – Duration of activities for Neil Armstrong

The result of this query shows the details and duration of each event that Neil Armstrong participated in during the Apollo mission.

By using the same view, we can query to see which astronauts walked on the Moon during the Apollo 11 mission and how long they spent on the lunar surface:

```
SELECT *
FROM apollo_activities
WHERE astronaut_location = 'Moon'
ORDER BY event_time;
```

We get the following result:

event_description varchar	event_time timestamp	astronaut varchar	astronaut_location varchar	end_time timestamp	event_duration interval
First step on lunar surface, That's one small st…	1969-07-21 02:56:00	Neil Armstrong	Moon	1969-07-21 05:09:00	02:13:00
Buzz Aldrin on lunar surface	1969-07-21 03:15:00	Buzz Aldrin	Moon	1969-07-21 05:01:00	01:46:00

Figure 4.18 – Duration of time spent on the Moon

This shows that Neil Armstrong spent 2 hours and 13 minutes on the lunar surface, and Buzz Aldrin spent 1 hour and 46 minutes on the lunar surface.

The LEAD and LAG SQL functions are particularly useful for analyzing data over time in DuckDB. Our Apollo 11 dataset allows us to access timestamps from the previous or following rows. This single-pass approach allows us to aggregate complex datasets without the need for multiple queries using self-joins.

Summary

In this chapter, we learned about DuckDB operations and performance. We explored the features and techniques that can help optimize the speed of DuckDB. We used indexes to accelerate queries and explored effective ways to manage and examine file operations on Parquet data files.

We also explored timestamps, seeing how DuckDB represents points in time internally and how it bins them into structured timestamps, as well as the usage of time zone information. We also looked at how window functions can be used to effectively process time series data.

Our next chapter introduces some powerful extensions that are available in DuckDB and how they can be used to extend DuckDB's capabilities across a range of specialized applications.

DuckDB Extensions

In this chapter, we explore DuckDB extensions, which provide a flexible way to add specialized functionality to DuckDB and your workflows. We will review the steps for installing and loading DuckDB extensions and go through some of the powerful capabilities that lie outside the core functionality of DuckDB. We will perform efficient queries on large collections of text data, explore data from remote systems, connect to data in other databases, and jump into the fascinating world of geospatial data processing.

In this chapter, we're going to cover the following main topics:

- Expanding DuckDB capabilities by exploring some of its extensions
- Working with geospatial data using DuckDB's `spatial` extension

Technical requirements

The files and example data for these exercises are available at `https://github.com/PacktPublishing/Getting-Started-with-DuckDB/tree/main/chapter_05`.

Expanding DuckDB capabilities with extensions

We have explored a range of functionalities that DuckDB provides out of the box for data exploration and analysis. DuckDB also supports extensions, which provide a mechanism for adding additional features that enhance its functionality.

DuckDB provides a range of officially supported extensions, with commonly used ones being installed and enabled by default in most DuckDB client distributions. Some extensions are also automatically installed and loaded the first time you make use of their functionality. You can readily install both the official extensions and third-party extensions from remote and local locations.

Extensions can provide a range of different types of functionality, often extending the supported data sources, data formats, data types, operations, or functions. For example, DuckDB's ability to work with JSON files is enabled by the `json` extension; the `spatial` extension adds support for geographic data types and spatial operations; the `postgres` extension provides a scanner that enables DuckDB to read directly from running PostgreSQL database instances. Let's dive in and explore some of the popular extensions that DuckDB makes available.

Available extensions in DuckDB

You can check the list of available DuckDB extensions along with their installed and loaded status by using the `duckdb_extensions` function. Let's look at a few interesting columns from this table function; however, note that the output may vary depending on which DuckDB version you have installed:

```
SELECT * FROM duckdb_extensions();
```

We can see that this has given a table of the available DuckDB extensions, along with information about each of them:

extension name varchar	loaded boolean	installed boolean	install path varchar	description varchar	aliases varchar[]	extension version varchar
arrow	false	false		A zero-copy data integration between Apache Arrow and DuckDB	[]	
autocomplete	true	true	(BUILT-IN)	Adds support for autocomplete in the shell	[]	
aws	false	false		Provides features that depend on the AWS SDK	[]	
azure	false	false		Adds a filesystem abstraction for Azure blob storage to DuckDB	[]	
excel	true	true	(BUILT-IN)	Adds support for Excel-like format strings	[]	
fts	true	true	(BUILT-IN)	Adds support for Full-Text Search Indexes	[]	
httpfs	false	false		Adds support for reading and writing files over a HTTP(S) connection	[http, https, s3]	
iceberg	false	false		Adds support for Apache Iceberg	[]	
icu	true	true	(BUILT-IN)	Adds support for time zones and collations using the ICU library	[]	
inet	true			Adds support for IP-related data types and functions	[]	
jemalloc	true	true	(BUILT-IN)	Overwrites system allocator with JEMalloc	[]	
json	true	true	(BUILT-IN)	Adds support for JSON operations	[]	
motherduck	false	false		Enables motherduck integration with the system	[md]	
mysql scanner	false	false		Adds support for connecting to a MySQL database	[mysql]	
parquet	true	true	(BUILT-IN)	Adds support for reading and writing parquet files	[]	
postgres_scanner	false	false		Adds support for connecting to a Postgres database	[postgres]	
shell	true				[]	
spatial	false	false		Geospatial extension that adds support for working with spatial data and…	[]	
sqlite_scanner	false	false		Adds support for reading and writing SQLite database files	[sqlite, sqlite3]	
substrait	false	false		Adds support for the Substrait integration	[]	
tpcds	false	false		Adds TPC-DS data generation and query support	[]	
tpch	true	true	(BUILT-IN)	Adds TPC-H data generation and query support	[]	
22 rows						7 columns

Figure 5.1 – DuckDB's catalog of extensions – an overview

This output corresponds to the default extension configuration in this DuckDB CLI session before any further extensions have been installed or loaded. Some columns are worth highlighting:

- `loaded`: A Boolean value that indicates whether the extension has been loaded for the current DuckDB process. Note that an extension must be installed before it can be loaded.

- `installed`: A Boolean value that indicates whether the extension is installed in the DuckDB client in use. An extension either comes built into your DuckDB client, can sometimes be automatically installed by DuckDB when needed, or can be explicitly installed by a user.

- `install_path`: If an extension is installed, this value represents the full system path for the extension. The (BUILT-IN) designator indicates that the code for the extension has been bundled for the current DuckDB client.

- `aliases`: Alternative names for an extension, which can be used when installing and loading each extension.

In this chapter, we'll focus on exploring a selection of the official DuckDB extensions. Some of these do not come built-in with DuckDB and need to be downloaded and installed explicitly (the `sqlite` and `fts` extension). We also look at one extension (the `httpfs` extension), which doesn't come pre-installed but will be automatically installed and loaded by DuckDB when you first use it. Note that while we don't cover it in this book, it's also possible to use extensions developed by third parties as well as develop and compile your own DuckDB extensions. To develop your own DuckDB extension, have a look at the template guide at `https://github.com/duckdb/extension-template`.

The SQLite extension

The `sqlite` extension (also known as `sqlite_scanner`) allows DuckDB to read and write data from a database file created by **SQLite**. As we discussed in *Chapter 1*, SQLite is a widely used in-memory database that is frequently used in mobile applications and embedded applications. However, since it is designed for applications involving row-oriented transactions, it struggles with performance on columnar, analytical-style workloads. Using the `sqlite` extension allows us to bridge this gap by enabling DuckDB to read and write data from SQLite database files. Once connected to a SQLite database, you can use DuckDB to query directly from underlying tables in the SQLite database. This allows us to take advantage of DuckDB's powerful analytical capabilities when working with data stored in SQLite databases.

To illustrate this workflow, we have created an SQLite database file called `my_sqlite.db`, which is available in this chapter's GitHub resources. Let's now see how we can use the `sqlite` extension to read from this SQLite database using DuckDB.

We first need to install and load the `sqlite` extension, which we do by using the following two commands:

```
INSTALL sqlite;
LOAD sqlite;
```

Executing the INSTALL command causes DuckDB to download the sqlite extension and then install it into your local DuckDB installation. The LOAD command is needed to enable the extension for use in your current session. If we now query the duckdb_extensions function, we'll see the extension has been both installed and loaded. We should note that internally, the sqlite extension is referred to as sqlite_scanner, with the sqlite alias being the standard name that the DuckDB documentation refers to it by.

```
SELECT extension_name, installed, loaded
FROM duckdb_extensions()
WHERE extension_name = 'sqlite_scanner';
```

We get the following results:

extension_name varchar	installed boolean	loaded boolean
sqlite_scanner	true	true

Figure 5.2 – The status of the sqlite extension after installation

With the extension installed, DuckDB has now acquired the new ability to read an external SQLite database. We can open the my_sqlite.db SQLite database file using the ATTACH command:

```
ATTACH 'my_sqlite.db' (TYPE sqlite);
```

With the external SQLite database available within DuckDB, we can now use it in a query. Let's look at the countries_sqlite table, which is provided in the included SQLite database. Note that we are using the table schema prefix my_sqlite to instruct DuckDB to read the data from the attached SQLite database:

```
SELECT *
FROM my_sqlite.countries_sqlite;
```

We get the result as follows:

country varchar	name varchar
AD	Andorra
AE	United Arab Emirates
AF	Afghanistan
AG	Antigua and Barbuda
AI	Anguilla

Figure 5.3 – Sample of countries displayed

The ability to load extensions and include them using DuckDB SQL is a powerful technique to support complex analysis activities over data stored in SQLite databases, which do not natively provide good support for analytical workloads. Let's jump into our next DuckDB extension.

Reading and writing remote files using the HTTPFS extension

DuckDB's httpfs extension enables DuckDB to both read and write remote files. In *Chapter 3*, we saw an example of how we can use it to directly query hosted Parquet files over HTTPS connections. In addition to being able to read publicly hosted files over HTTP(S), the httpfs extension also enables the reading and writing of files in cloud object storage using the S3 API. Let's now revisit using this extension for querying public files over HTTP(S). We'll then go a bit further by querying files stored in an AWS S3 bucket.

Many useful datasets are openly available on the public internet. For example, the United States Census Bureau (https://www.census.gov) makes available data containing information on population, demographics, and housing.

Let's look at an example that involves querying a remotely-hosted CSV file containing city-level population data over HTTPS. We'll start by installing and loading the httpfs extension:

```
INSTALL httpfs;
LOAD httpfs;
```

Note that the httpfs extension is capable of being automatically installed and loaded. This means that you can skip the preceding steps and DuckDB will automatically install and load this extension the first time you issue a query or statement that makes use of it.

With the `httpfs` extension installed and loaded, we'll now directly query our target CSV file:

```
SELECT *
FROM read_csv('https://www2.census.gov/programs-surveys/popest/
datasets/2020-2022/cities/totals/sub-est2022.csv');
```

We get the following result:

NAME varchar	STNAME varchar	ESTIMATESBASE2020 int64
Alabama	Alabama	5024356
Abbeville city	Alabama	2355
Adamsville city	Alabama	4372
Addison town	Alabama	661
Akron town	Alabama	227

Figure 5.4 – Sample of city populations read from a remote file

As we can see, this query has retrieved the contents of the remotely-hosted CSV file `sub-est2022.csv` and has also parsed the results into a table.

We can also use the DuckDB `httpfs` extension to interact with files located on cloud-object storage using the S3 API. While it might sound like this refers specifically to the AWS S3 (Simple Storage Service) service, a number of other popular cloud storage services offer S3 API compatibility. At the time of writing, the `httpfs` extension has been tested against the following object storage services: **AWS S3**, Google Cloud's **Cloud Storage**, **lakeFS**, and **Minio**. Other services offering S3 API-compatible cloud storage should also work; however, it's possible some features may not be supported.

To demonstrate querying from cloud storage, we have created an AWS S3 bucket (`s3://duckdb-s3-bucket-public`) that contains a simple Parquet file: `countries.parquet` that maps country codes to country names.

In order to query from the files hosted in an S3-compatible object storage, we first need to configure the region. Setting the endpoint is optional; however, you will need to do this if you're working with an object storage that's not AWS S3.

As of DuckDB 0.10.0 and higher, the preferred way to configure S3 parameters, such as region and endpoint settings, is to use a DuckDB `SECRET`. We can create an S3 secret for this exercise like this:

```
CREATE OR REPLACE SECRET mysecret (
    TYPE S3,
    REGION 'us-east-1',
    ENDPOINT 's3.amazonaws.com'
);
```

By default, a secret is temporary and only held in memory, and it will vanish when the session is closed. Alternatively, a persistent secret is usable between DuckDB invocations. It is worth noting that persistent secrets are stored on disk at `~/.duckdb/stored_secrets` in an unencrypted binary format.

With our configuration completed, we can now directly query our Parquet file from its S3 bucket. For example, we can query the list of countries, searching for names that contain the string `Republic` with a query against the S3 bucket:

```
SELECT *
FROM read_parquet('s3://duckdb-s3-bucket-public/countries.parquet')
WHERE name SIMILAR TO '.*Republic.*';
```

The result of this query is as follows:

```
| country |         name             |
| varchar |         varchar          |
|---------|--------------------------|
| CF      | Central African Republic |
| CG      | Congo [Republic]         |
| CZ      | Czech Republic           |
| DO      | Dominican Republic       |
```

As you can see, we've successfully queried and returned the structured contents of the `countries.parquet` file, returning only the names that have the `Republic` string.

This bucket is public, meaning we didn't need to authenticate ourselves to get access to the file. If the file is not publicly accessible, the DuckDB `httpfs` extension can be configured to set an access key ID and secret access key. You can learn more about configuring secrets and the methods for handling multiple secrets for a variety of cloud service providers in the DuckDB documentation (`https://duckdb.org/docs/sql/statements/create_secret`).

Once you have set these with appropriate credentials, not only will you be able to read from cloud storage that requires authentication, but you'll also be able to write files to cloud storage—provided, of course, you have the appropriate permissions to do so.

Now that we have seen the use of the `httpfs` extension to read remote files, let's move onto exploring another DuckDB extension to help us plan for a holiday.

Full-text search indexes

Let's now look at the **Full-Text Search** (fts) DuckDB extension, which adds support for searching over textual data stored in DuckDB. Using this extension involves the creation of full-text search indexes that enable the efficient querying of large text collections stored as columns in DuckDB tables. More generally, this feature of a database is commonly referred to as **full-text search**, which is where the extension gets its name. Among other applications, full-text search indexes can be used to build search functionality that enables users to find relevant information based on natural language queries composed of words or phrases.

In order to jump into using DuckDB's full-text search functionality, let's imagine a situation where we want to build a search feature on top of a database that contains catalogue information about a collection of travel books. Our motivating use case will be someone with an interest in wine, who is planning a holiday to France. We want to enable them to make targeted queries of the database around their travel interest to find books that mention specific words or phrases within their product information.

To do this, we'll use the fts extension to create indexes on the columns containing each book's title and description. This will enable the user to perform fast queries across the entire collection. If you haven't come across full-text indexes previously, you may wonder why we can't just use matching string values using regular expressions or DuckDB's LIKE operator, which provides another type of string matching that uses simpler patterns. This approach has the practical deficiency of not being able to scale beyond rather small collections, as it requires every value in a column to be scanned at query time. Hence, there is a need for a specialized index that is able to map search terms in a query to documents that contain them without the need to scan every document's full contents at query time. Additionally, a specialized index will also offer us some control over how text is split up into component terms that are inserted into the index, which allows us to fine-tune search performance for specific applications.

For this exercise, we first need to load some data. We'll be using the book review data found in the external book reviews dataset we used in *Chapter 4*, so make sure you have the file books_data. csv ready to go. If you don't have it, refer to the directions in the *Downloading the book review dataset from Kaggle* section from *Chapter 4*.

Before we can create our table that contains book records, we need to address the lack of an ID field in this dataset. We'll do this using DuckDB's CREATE SEQUENCE statement. You may recall from *Chapter 4* that a sequence is used to create a unique ascending number. Let's create a sequence that we'll use to add a unique ID to each book record:

```
CREATE OR REPLACE SEQUENCE book_details_seq;
```

With the sequence created, let's load the book data into a new table called book_details:

```
CREATE OR REPLACE TABLE book_details AS
SELECT nextval('book_details_seq') AS book_details_id,
    Title AS book_title,
```

```
    description AS book_description
FROM read_csv('../chapter_04/books_data.csv');
```

We've now created a table that contains each book in our collection, along with columns for the book's ID, its name, and a description of its contents. Let's get a sense of what's inside this dataset by summarizing the contents of the book_details table:

```
SUMMARIZE book_details;
```

We get the following result:

column_name varchar	column_type varchar	min varchar	...	count int64
book_details_id book_title book_description	BIGINT VARCHAR VARCHAR	1 " Film technique, ... !! ALL NEW CAMPGRO.	... ...	212404 212404 212404

Figure 5.5 – Columns and sample values from the book_details table

Great work; we now have over 200,000 book records loaded into the table book_details. Let's jump into using the Full-Text Search extension to efficiently search through this large collection.

We'll start by installing and loading the fts extension:

```
INSTALL fts;
LOAD fts;
```

We want to create a full-text search index for the data in the book_title and book_description columns to allow us to perform fast and flexible searches on the text fields. Now, we'll use the create_fts_index pragma command to create a full-text search index on our target columns:

```
PRAGMA create_fts_index(
    'book_details',
    'book_details_id',
    'book_title',
    'book_description',
    overwrite=true
);
```

Once that's completed, we'll have a full-text search index that can be used to query the `book_title` and `book_description` columns. We'll make queries against the index by using a retrieval macro called `match_bm25`, which is created automatically when the `create_fts_index` function is invoked. This macro is created under the schema that follows the pattern `fts_<schema>_<table-name>`. Since our `book_details` table is located in the `main` schema, our fully qualified macro name is `fts_main_book_details.match_bm25`, which is how we'll need to invoke it.

With our index created, let's try it out with a travel-related query against our book collection made by our wine-loving user, who is planning their trip to France. Let's find a book for which the title or description includes any of the words `travel`, `france`, and `wine`. We will use a **common table expression** (**CTE**) to build our query. You may recall from the *Common table expressions* section in *Chapter 3* that a CTE can be thought of as an inline view that captures a modular component of query logic:

```
WITH book_cte AS (
    SELECT *,
        fts_main_book_details.match_bm25(
            book_details_id,
            'travel france wine'
        ) AS match_score
    FROM book_details
)
SELECT book_title, book_description, match_score
FROM book_cte
WHERE match_score IS NOT NULL
ORDER BY match_score DESC
LIMIT 10;
```

The full query first creates a CTE named `book_cte`, which performs our desired full-text search by retrieving every book record in our collection and augmenting it with a corresponding score for our target search terms, which is generated using the `fts_main_book_details.match_bm25` macro. We then query this, selecting from `book_cte` and retrieving product information for the books with the 10 highest-matching scores:

book_title varchar	book_description varchar	match_score double
Passions : The Win…	This is a biography of Thomas Jefferson at leisure, enjoying two of his passions--wine and travel	9.61449186708193
France: Best Place…	Offers an illustrated guide to the wines and foods of Frances sixteen regions, along with informa	8.91698455650084
A Travellers Wine…	Italy has probably the richest variety of wines of any country in the world. Stephen Hobley descr	7.260535148321603
Rhone Renaissance:…	Norman assesses more than 1,500 Rhone-style wines from over 220 estates on four continents. Such	7.002791050402966
The Judgement of P…	Looks at an event held in 1976 in which French judges, during a blind taste-test, chose unknown C	6.96018856787726
The Rough Guide to…	Now available in ePub format. The Rough Guide to France is the ultimate travel guide to this vari	6.878193972053833
The French Vineyar…	A Michelin three-star French chef conducts readers on an illustrated tour of the wine-making regi	6.838477037639243
The Best 50 Bargai…	The most knowledgeable wine drinkers know that there are inexpensive wine bargains to be found, f	6.767250622438753
Women of Wine: The…	This book, with its personal approach and global scope, is the first to explore womens increasing	6.473852734568955
Africa Uncorked: T…		6.3959210810579386

Figure 5.6 – Ten books with the highest matching scores – an overview

As you might expect, the match score will be higher for books that contain more instances of our search terms in their title or description. By default, the query does not require all terms to be present within a document in the index. This means that some of our preceding matching records might feature only one or two of the terms `travel`, `france`, and `wine`. We can require all terms to be present in matching fields by setting the `conjunctive:=1` parameter of the `match_bm25` macro. If you re-run the preceding query with this parameter set, you should find fewer results returned, which are the records for which the indexed document did not contain all terms. It is also possible to specify the parameters of the `create_fts_index` function, allowing you to customize the preprocessing steps that are applied to text to convert them into terms (or **tokens**) that are inserted into the index. This includes customizations such as disabling conversion to lowercase text, disabling the stripping of character accents, and controlling which words are excluded from being indexed (referred to as **stopwords**, which are words such as *is* and *the*, which can be disregarded from the index). See the `fts` extensions documentation page (`https://duckdb.org/docs/extensions/full_text_search`) for a complete list of the customizations that you can make to index creation and querying.

It looks like our full-text search index has successfully enabled our user to find some helpful book suggestions to help plan their upcoming trip to some French wine regions. In the next section, we'll continue with the theme of planning this holiday; this time, we will use DuckDB's `spatial` extension to help plan our trip.

Working with geospatial data

Geospatial data describes the location and attributes of spatial objects. The DuckDB `spatial` extension adds support for working with geospatial data types. With the `spatial` extension loaded, we can store, query, analyze, and manipulate geospatial data in DuckDB using SQL commands.

The primary data type that the spatial extension introduces is the `GEOMETRY` data type. Underlying each `GEOMETRY` object is a more specific geometric subtype, such as `POINT` and `LINE`, which we will see shortly as part of researching a trip to the wine regions of France. We'll be using geospatial data to calculate the distances between locations, and we'll also be applying spatial filters to find candidate wine destinations to visit.

The spatial extension

Our first step is to install and load the DuckDB `spatial` extension:

```
INSTALL spatial;
LOAD spatial;
```

With the extension installed and loaded, let's start our exploration of Paris. Imagine we are next to the Eiffel Tower. We could use a handheld GPS device to pinpoint our location:

- Latitude: 48.858935

- Longitude: 2.293412

Let's start by creating a DuckDB object that represents the location of the Eiffel Tower. The `spatial` extension provides the `st_point` function, which creates a `POINT` geometry object from two `DOUBLE` values, representing a point in a geographic co-ordinate system:

```
SELECT st_point(48.858935, 2.293412) AS Eiffel_Tower;
```

We get the following result:

Eiffel_Tower geometry
POINT (48.858935 2.293412)

Figure 5.7 – The GPS location of the Eiffel Tower

We have created a geometry point for the Eiffel Tower; however, we have yet to define which spatial reference system we're working with. The two numbers provided to the `st_point` function could be any x, y co-ordinates in any number of different co-ordinate reference systems. Like many geographic information systems, the DuckDB `spatial` extension allows us to use **EPSG** codes that define spatial reference systems for representing points on the Earth. The acronym EPSG has its origins in the **European Petroleum Survey Group,** with the EPSG code uniquely identifying which co-ordinate system or spatial reference system is in use.

Different spatial reference systems correspond to different strategies for projecting naive two-dimensional co-ordinates onto the spherical surface of the Earth. Each projection strategy has different trade-offs with respect to what type of distortions are introduced from the projection process, which means that it can be important to choose one that is suitable for your application.

For this exercise, we will be using two EPSG spatial reference systems:

- `EPSG:4326`, which is used for representing geographic co-ordinates as latitude and longitude pairs, with degrees as the unit

- `EPSG:27563`, which is a projection that covers northern France and uses meters as the unit

Now that we've covered some of the fundamentals, let's see if we can combine a few DuckDB spatial commands to work out the distance between two points. Given that we're currently standing at the foot of the Eiffel Tower, let's work out how far away the Arc de Triomphe is:

```
SELECT
    st_point(48.858935, 2.293412) AS Eiffel_Tower,
    st_point(48.873407, 2.295471) AS Arc_de_Triomphe,
    st_distance(
        st_transform(Eiffel_Tower, 'EPSG:4326', 'EPSG:27563'),
        st_transform(Arc_de_Triomphe, 'EPSG:4326', 'EPSG:27563')
    ) AS Aerial_Distance_M;
```

There's a bit going on with this SQL query. We first created two points that represent the location of the Eiffel Tower and the Arc de Triomphe using the st_point function. We then used the st_transform function to transform these points from the EPSG:4326 latitude and longitude degrees to the EPSG:27563 co-ordinate system (which is based on meters and covers northern France). Finally, we used the st_distance function to calculate the distance between these two points in meters:

Eiffel_Tower geometry	Arc_de_Triomphe geometry	Aerial_Distance_M double
POINT (48.858935 2.293412)	POINT (48.873407 2.295471)	1622.0284586410457

Figure 5.8 – Measuring the distance between the Eiffel Tower and the Arc de Triomphe

We can see that there are 1,622 meters (as the straight-line distance) between the Eiffel Tower and the Arc de Triomphe, which could be a nice afternoon walk.

Reading Microsoft Excel files

A perhaps surprising feature of the DuckDB spatial extension is its ability to read and write the xlsx file format. The Microsoft Excel Open XML Spreadsheet format is the default format used by Microsoft Excel for saving spreadsheets authored with the tool.

We can use the st_read function included within the spatial extension to read Microsoft Excel files. Let's look at the names and locations of French train stations that have been stored in an Excel file called stations.xlsx. The data we want are located in a sheet named layers, which we can specify using the layer parameter of the st_read function:

```
SELECT *
FROM st_read('stations.xlsx', layer='stations');
```

Running this query gives us the following output:

station_name varchar	longitude double	latitude double
Abancourt	1.774306	49.685224
Abbaretz	-1.524416	47.554643
Abbeville	1.82449	50.10221
Abbémont	2.50808	49.616695
Aboncourt	6.35	49.266667

Figure 5.9 – Sample of rows from the stations Excel file

We can see that the st_read function has read the train station location data from the Excel file and determined the appropriate data types for the columns. Let's now create a table for this data:

```
CREATE OR REPLACE TABLE stations AS
SELECT *
FROM st_read('stations.xlsx', layer='stations');
```

We have now created a stations table, containing the records for each station that include its name and the latitude and longitude co-ordinates. The ability to read Microsoft Excel files is a useful and somewhat hidden capability of the DuckDB spatial extension. We'll use this table in the next section to locate train stations in a particular region.

Points within a polygon

We've seen how to calculate the distance between two points, but let's try a more interesting example using the DuckDB geometry data types.

Imagine we wish to plan a trip to a famous wine-producing region of France. The Bordeaux wine region is one of the most famous and prestigious wine-producing areas in the world. The boundary of the Bordeaux wine region is strictly defined, with only wines from a very specific geographical area being eligible to be considered a Bordeaux wine. The perimeter of this wine region is a complex shape, so it's a great opportunity to load some data into DuckDB that defines the geometry of this border and then make use of this to help with planning our holiday.

We can use the POLYGON spatial type to store the list of geographic features, such as the boundary markers of the Bordeaux wine region. We can use the geojson file format to encode the co-ordinate pairs that specify the vertices of POLYGON that encompass the area of interest.

The bordeaux_wine_region.geojson file represents the geographic boundary of the Bordeaux wine region. We can load this file into DuckDB using the st_read function:

```
SELECT geom
FROM st_read('bordeaux_wine_region.geojson');
```

This query gives us the following result:

geom
geometry
POLYGON ((-1.1535645 45.5756002, -0.6481934 44.6022017, -0.1963806 44.5053207, -0.1922607 44..

Figure 5.10 – Sample of points using the Bordeaux wine region

We can see that this query returns a single record containing a POLYGON object. This polygon contains a list of coordinate points that represent the boundaries of the area that contains our next holiday destination. Recall that we have loaded all French train station records into the stations table, which contains station latitude and longitude co-ordinates. Let's see if we can find the train stations within the perimeter of the Bordeaux wine region. We will use the st_within function to test if a station location falls within the boundary. The st_within function takes two arguments, with each being a GEOMETRY subtype. It tests whether the first argument's value falls inside the second. For our context, we want to test to see whether a POINT (representing a train station) falls inside a POLYGON (representing the boundaries of the Bordeaux region):

```
SELECT station_name
FROM stations
WHERE st_within(st_point(longitude, latitude),
    (
        SELECT  geom
        FROM st_read('bordeaux_wine_region.geojson')
    )
);
```

Here's the result:

station_name varchar
Alouette France
Arbanats
Aubie-St-Antoine
Barsac
Bassens
Beautiran
Blanquefort
Bordeaux
Bordeaux St-Jean

Figure 5.11 – Sample of train station names

The query returns the names of the 61 train stations that are located within the Bordeaux wine region. The `st_within` function within the WHERE clause serves as a predicate, checking if each of the points representing the stations is completely inside the polygon that defines the Bordeaux region. The first argument to the function is a POINT value created using `st_within`, and the second argument is a POLYGON value returned using a SELECT statement that converts the WKB data from the GeoJSON file into a GEOMETRY object. Finding points within a polygon is a powerful technique, and we could, for example, use a similar approach to find buildings at risk of flooding or finding restaurants within a popular district.

This set of examples has helped illustrate how DuckDB's `spatial` extension enhances DuckDB's capabilities by supporting geospatial data types and operations. The ability to store, query, and analyze geospatial data is a particularly valuable addition to the analytical capabilities of DuckDB.

Summary

In this chapter, we saw how DuckDB extensions enhance it with a range of specialized features and capabilities that are not available in DuckDB core. We saw how these extensions enable DuckDB to directly query other database systems such as SQLite, read and write remotely hosted files, and perform powerful and efficient full-text searches over large collections of documents. We also went on a deeper dive into the `spatial` extension, seeing how we can leverage the geospatial data types and operations it provides for performing analytical queries on geospatial data.

You should now feel comfortable with the process of installing and loading DuckDB extensions, which will allow you to leverage their powerful capabilities in your own workflows and projects. We should also highlight that we have only explored a selection of DuckDB's extensions in this chapter. We encourage you to explore DuckDB's full catalogue of extensions, to see what other functionality to can take advantage of: `https://duckdb.org/docs/extensions`.

In the next chapter, we'll be departing from our exploration of DuckDB's features via its CLI client and moving on to exploring DuckDB's Python client. This DuckDB client offers a particularly compelling integration of DuckDB into rich data analysis workflows as well as into operational analytical data infrastructure and data applications.

6

Semi-Structured Data Manipulation

Data is generated by a wide range of systems, in an endless variety of shapes. Sometimes, the data you'll need to work with won't conform to the tabular structure of rows and columns that we tend to associate with relational databases. **Semi-structured data** refers to data that does not follow a strict tabular schema or data model. Such data may feature some or all the following properties:

- Composite data structures involving multiple values within a single entry, such as sequences of values and groupings of named values within a single entry

- Flexible schemas, both within composite values and in the shape of data across records

- Nested data, where composite values contain further composite values, enabling the modeling of hierarchical data within individual values

A particularly well-known form of semi-structured data is **JavaScript Object Notation (JSON)**, which you will almost certainly find yourself having to wrangle at one point or another due to its ubiquity as a data interchange format that's used in all manner of systems, such as web **application programming interfaces (APIs)**, **Internet of Things (IoT)** devices, web browsers, and mobile apps.

Analytical workloads frequently involve integrating diverse data sources, something that requires synthesizing and managing heterogeneous schemas. Many of these data sources also feature semi-structured data themselves, such as web APIs, cloud services, logging and monitoring data, and IoT devices. DuckDB offers a range of data types, built-in functions, and SQL enhancements for effectively working with semi-structured data. In this chapter, we'll explore a selection of DuckDB's features for modeling, manipulating, and querying semi-structured data. In particular, we'll cover the following topics:

- Exploring some of DuckDB's nested data types, including the LIST, MAP, and STRUCT data types

- Working with the LIST data type

- Working with JSON data, using DuckDB's json extension
- Working with JSON data returned from APIs

Technical requirements

In this chapter, we need to get data into DuckDB so that it can be queried, transformed, and analyzed. The files and example data for these exercises are available in this book's GitHub repository: https://github.com/PacktPublishing/Getting-Started-with-DuckDB/tree/main/chapter_06.

This chapter also makes use of the TVmaze API (https://www.tvmaze.com/api), which is licensed under CC BY-SA (http://creativecommons.org/licenses/by-sa/4.0/).

Exploring nested data types

DuckDB offers a range of nested, or composite, data types. In contrast to single-valued data types such as INTEGER and VARCHAR, nested data types are characterized by being able to store multiple values in the form of sequences of values and structures containing multiple named-property values. These nested types can contain further nested types, allowing hierarchical data to be modeled within each row of a table. This makes them a natural fit for modeling different kinds of semi-structured data within DuckDB. In this section, we'll introduce a selection of DuckDB's nested data types: LIST, MAP, and STRUCT.

The LIST data type

A DuckDB LIST is an ordered sequence of values that all have the same data type. When used as the type of a column, each row must share the same data type for each LIST instance; however, lists can have variable lengths across rows. A LIST literal is represented as a comma-separated list of values, wrapped in square brackets. Here's an example:

```
SELECT [7,8,9] AS list_int;
```

This creates a LIST data type of INTEGER values that contains our three numbers. When we look at the result of this query, DuckDB renders this data type as int32[]. The square brackets indicate we are dealing with a list of values that are all the same type:

list_int int32[]
[7, 8, 9]

Figure 6.1 – The result of creating an integer LIST data type

LIST can be empty (with zero elements) and is essentially unbounded, allowing for any number of elements. However, note that every element within LIST must have the same data type. You cannot mix INTEGER and VARCHAR values in a LIST data type, for example.

Now, let's create a LIST data type of VARCHAR values that represent the names of recent James Bond movies:

```
SELECT [
     'Quantum of Solace',
     'Skyfall',
     'Spectre',
     'No Time to Die'] AS list_string;
```

This returns a LIST data type containing VARCHAR values that DuckDB has rendered as varchar[]:

list_string varchar[]
[Quantum of Solace, Skyfall, Spectre, No Time to Die]

Figure 6.2 – The result of creating a LIST value

We'll also mention another nested DuckDB data type that is similar to the LIST data type: the ARRAY data type. The primary difference between them is that when used as the type of a column, LIST values can store variable-length sequences across rows, whereas ARRAY values must have a fixed length across each row. This fixed size enables DuckDB to optimize how ARRAY values are stored and retrieved, which makes them well-suited for efficiently storing and retrieving numerical vectors. They are frequently used in machine learning and data science applications.

Next, we'll look at the MAP data type, which offers some of the flexibility of the LIST data type in that it supports columns containing nested elements of variable length but differs in that its nested elements take the form of named properties.

The MAP data type

The MAP data type provides an ordered sequence of elements, each mapping a key to a corresponding value. Conceptually, you can think of a MAP value as being composed of two lists – one for the keys and one for the values. When creating MAP values, we need to be consistent with our data types. All the keys within a MAP instance must be of the same data type. Equally, we need to be consistent and use the same data type across all its values.

For example, let's create a table containing a single MAP value that links keys representing movie names (VARCHAR values) to their year of theatrical release (INTEGER values):

```
CREATE OR REPLACE TABLE movies AS
SELECT MAP(
    [
        'Quantum of Solace',
        'Skyfall',
        'Spectre',
        'No Time to Die'
    ],
    [2008, 2012, 2015, 2021]
) AS movie_release_map;
```

Before we see how to retrieve values from MAP, let's see how the MAP object appears when we query it from our table:

```
SELECT movie_release_map
FROM movies;
```

This query retrieves the single row in our table, showing us that the table has a single column whose data type is a MAP value that associates VARCHAR values with INTEGER values. In this single MAP value, we can see that it contains each of the key-value pairs we defined, linking movie names to their year of release:

movie_release_map map(varchar, integer)
{Quantum of Solace=2008, Skyfall=2012, Spectre=2015, No Time to Die=2021}

Figure 6.3 – The result of creating a MAP data type

With this table created, let's see how values can be retrieved from inside our MAP value. We can retrieve the value corresponding to a specific key within a MAP value by specifying the desired key in square brackets. For example, we can retrieve the release date for 'Quantum of Solace' with a query like this:

```
SELECT movie_release_map['Quantum of Solace']
FROM movies;
```

This retrieves the release date of 2008:

```
movie_release_map['Quantum of Solace']
                 int32[]
```
```
[2008]
```

Figure 6.4 – The result of retrieving a MAP value

Note that selecting from a MAP value always returns a LIST data type containing the value corresponding to the queried key, or an empty LIST if the key was not found in the MAP value. If you want to retrieve the underlying value directly, you'll need to use LIST selection syntax, like this:

```
SELECT movie_release_map['Quantum of Solace'][1]
FROM movies;
```

Next, we'll look at the STRUCT data type, which, like the MAP data type, allows us to store named property fields within a single column but is better suited to modeling data when each record follows a uniform schema.

The STRUCT data type

STRUCT is a composite data type with a fixed number of ordered *entries*, each of which has a name and corresponding value. The values within a STRUCT data type can be retrieved by referencing the desired entry name, referred to as its *key*. When used as the type of a column, a STRUCT data type effectively allows you to nest multiple columns (the entries of STRUCT) inside a single column. It is important to note that across each row of a STRUCT column, the schema for the entries must be consistent across each row in the table, meaning they must all have the same key names and corresponding data types. The STRUCT data type enables us to group repeating data types together under a single structure. Due to their fixed shape and consistent data representation, DuckDB is extremely efficient at processing data stored within STRUCT columns.

While the keys of a STRUCT data type are always strings, a given STRUCT can contain a variety of data types. In addition to single-valued data types, such as INTEGER and VARCHAR, you can also place nested data types inside of STRUCT, such as LIST and MAP, and even other STRUCT values. Remember that while a given STRUCT can contain values of different types, all rows within a STRUCT column must adhere to the same schema.

Now, let's see STRUCT in action. We'll create a STRUCT data type with columns corresponding to the movie's name, release year, and box-office revenue in millions of USD:

```
SELECT {
    movie: 'No Time to Die',
    release_year: 2021,
    box_office: 771.2
} AS struct_movie;
```

This gives us the following result:

struct_movie
struct(movie varchar, release_year integer, box_office decimal(4,1))
{'movie': No Time to Die, 'release_year': 2021, 'box_office': 771.2}

Figure 6.5 – A single STRUCT record that contains the properties of a movie

We can see that this query has created a single STRUCT column whose respective entries have been inferred as having the data types of VARCHAR and INTEGER, as well as a DECIMAL value that can store four digits to the left of the decimal point and one after.

STRUCT versus MAP

The STRUCT and MAP data types are conceptually similar in that they both contain ordered entries of keys mapping to values. Given this, you might be wondering when to use one over the other. When using a STRUCT column, each row must share the same schema, meaning that each STRUCT must have the same key present for each row. MAP columns, on the other hand, can have different keys present across each row. This makes MAP columns particularly useful when you don't know the schema in advance, or you know the schema will vary across records. This flexibility is the primary reason to use MAP over STRUCT. If you know that each row will have the same schema, then you should opt for using STRUCT as it confers more optimized storage and retrieval.

Another difference is that the keys of a STRUCT column are always strings. In contrast, MAP allows other data types to be keys, but all keys in MAP must share the *same* data type. The values within MAP also must share the same data type, which contrasts with STRUCT, whose values can have heterogeneous data types (but must be consistent across rows). Conceptually, it is helpful to think of MAP as being made of a list of keys of the same type that's aligned with a list of values of the same type.

Before we wrap up this section on DuckDB's nested data types, we'll mention the two DuckDB nested data types we have not covered here: ARRAY and UNION. As we mentioned earlier, the ARRAY data type stores fixed-length arrays of values, making it well-suited for optimized storage and retrieval of numerical vectors for data science applications. The UNION data type allows you to define columns whose single values can vary from a schema of possible named values, allowing you to store alternating data types across rows. We encourage you to explore these in the DuckDB nested type documentation, where you can also dive deeper into the features of the LIST, MAP, and STRUCT data types that we did not cover here.

This brings us to the end of our brief tour of a selection of DuckDB's nested data types, where we covered LIST, MAP, and STRUCT. We selected these data types to illustrate DuckDB's strength at modeling different flavors of commonly occurring semi-structured data that you might find in the wild. We'll be building on these foundations throughout this chapter, first when we dive deeper into working with the LIST data type in the next section, and later, when we cover how DuckDB allows us to effectively load, wrangle, and model data stored in JSON format.

Working with lists

In this section, we're going to dive a bit deeper into working with the LIST data type, which we introduced earlier in this chapter. DuckDB provides a range of features and affordances that make this a particularly powerful and flexible data type, some of which take inspiration from other languages, such as Python. Before we dive into playing with LIST types, we'll prepare some data that we'll use for this section, and which we'll also use later in this chapter.

Preparing the data

The data we'll be using consists of a small dataset of films, leading actors, and roles they play and is located in the film_actors.csv CSV file in this chapter's directory in this book's GitHub repository. We'll create a table called film_actors, which we'll import the contents of the CSV file into:

```
CREATE OR REPLACE TABLE film_actors AS
SELECT *
FROM read_csv('film_actors.csv');
```

Let's have a look at the contents of this dataset:

```
SELECT *
FROM film_actors;
```

This shows us the records in our small dataset of actors and their movie appearances:

film_name varchar	actor_name varchar	character_name varchar
James Bond - No Time to Die James Bond - No Time to Die James Bond - Spectre James Bond - Spectre James Bond - Spectre Barbie Barbie	Daniel Craig Ana de Armas Daniel Craig Léa Seydoux Christoph Waltz Margot Robbie Ryan Gosling	James Bond Paloma James Bond Madeleine Swann Blofeld Barbie Ken

Figure 6.6 – A dataset of films, actors, and the characters they played

Now that we've prepared the data, let's dive into working with LIST data types using this dataset.

List aggregation

A common task for data manipulation is aggregating many rows together into a smaller number of rows by grouping rows and combining each group into a single row. Aggregate functions such as SUM, MAX, and COUNT will be familiar to many SQL users for analytical applications that involve deriving numerical summary statistics of fields for each group of rows. Aggregate functions don't have to reduce rows into numerical values, however. A particularly useful pattern that DuckDB supports is aggregating column-wise row groups into LIST instances that contain the contents of each group as a single value. This functionality is enabled through DuckDB's list aggregate function.

Let's start with a simple aggregation, where we'll collapse the values occurring in the actor_name column into a single row that contains a single LIST data type whose elements are all our actor names as VARCHAR values. To apply this aggregation, we simply need to apply the list aggregate function to the actor_name column in our SELECT clause. Since we want to create a single row for the entire dataset, rather than multiple rows for subgroups, we don't need to include a GROUP BY clause for our query. Here's our query:

```
SELECT list(actor_name) AS actors
FROM film_actors;
```

This query gives us results in a single row with a LIST data type that contains every actor-name occurrence across all our movie records:

actors varchar[]
[Daniel Craig, Ana de Armas, Daniel Craig, Léa Seydoux, Christoph Waltz, Margot Robbie, Ryan Gosling]

Figure 6.7 – A LIST data type containing every actor-name occurrence in our dataset

Now, let's look at a query where we want to perform a list aggregation that results in multiple rows that correspond to groupings of interest. Let's say we wanted to retrieve the list of actors that appear in each movie in our dataset. This will be similar to our previous query as we'll apply the list aggregate function to the actor_name column again. However, this time, we'll also use a GROUP BY clause to perform our aggregation across groups of records with the same film_name. We'll also use an ORDER BY clause to provide an explicit ordering for results by film_name. Here's our new query:

```
SELECT film_name, list(actor_name) AS actor_name_list
FROM film_actors
GROUP BY film_name
ORDER BY film_name;
```

Looking at the results of this query, we can see that we've now got a record for each movie, which consists of its name, along with a LIST column of VARCHAR values containing the actors that appear in each movie:

film_name varchar	actor_name_list varchar[]
Barbie James Bond - No Time to Die James Bond - Spectre	[Margot Robbie, Ryan Gosling] [Daniel Craig, Ana de Armas] [Daniel Craig, Léa Seydoux, Christoph Waltz]

Figure 6.8 – Movies and lists of their starring actors

Notice that our actor_name_list column has variable-length LIST values. This is one of the key benefits of the LIST data type – it allows us to model this kind of semi-structured data.

Once you have some data containing LIST values, DuckDB offers a range of convenient and powerful features for working with them. We'll reuse both of these aggregation queries, which returned tables containing LIST data types, to explore some of the kinds of things we can do with LIST in DuckDB.

List functions

Having used the `list` aggregate function to create data containing a `LIST` data type, let's have a look at some of the scalar list functions that DuckDB provides – that is, functions that operate over `LIST` objects and produce a single row for each invocation, rather than changing the cardinality of the results, as we saw with list aggregation.

Returning to our example where we derived the list of actors occurring in each movie, let's say we wanted to return only two actors occurring in the second and third positions of each list. To do this, we can use the `list_slice` function, which extracts a slice of the elements in a list by positional indexes, which we pass as arguments. In this case, we want to slice from position 2 to position 3:

```
SELECT film_name,
    list_slice(list(actor_name), 2, 3) AS other_actors
FROM film_actors
GROUP BY film_name
ORDER BY film_name;
```

Looking at the results of this query, we can see that the list of actors for each movie has been reduced in size so that it only includes actors occurring in the second and third positions. Note that for movies that only have two actors listed in our dataset, there is no third position, so we get lists containing just a single actor, corresponding to the one in the second position:

film_name varchar	other_actors varchar[]
Barbie James Bond - No Time to Die James Bond - Spectre	[Ryan Gosling] [Ana de Armas] [Léa Seydoux, Christoph Waltz]

Figure 6.9 – Movies and sliced lists of their starring actors

DuckDB also provides a convenient syntax for slicing lists, which you might find familiar if you have used list slicing in Python. The following example is functionally equivalent to the previous SQL query:

```
SELECT film_name, list(actor_name)[2:3] AS other_actors
FROM film_actors
GROUP BY film_name
ORDER BY film_name;
```

One key difference from Python list indexing that you might have picked up on already if you have Python experience is that DuckDB counts its positional indexes starting at 1, whereas Python uses 0-based indexing.

Returning to the aggregation example where we created a single list containing every actor occurrence in the dataset, you might have noticed that this list contained duplicate actor names for those starring in multiple movies in the dataset. We can use DuckDB's `list_distinct` function to remove duplicate values from our list:

```
SELECT list_distinct(list(actor_name)) AS actors
FROM film_actors;
```

This query returns a list that's had duplicate actor names removed:

actors varchar[]
[Ana de Armas, Daniel Craig, Léa Seydoux, Christoph Waltz, Margot Robbie, Ryan Gosling]

Figure 6.10 – A list containing every distinct actor name in our dataset

Note that you might see a different ordering of names in your output, as the `list_distinct` function does not preserve the order of the original list. We can enforce an ordering using the `list_sort` function:

```
SELECT list_sort(list_distinct(list(actor_name))) AS actors
FROM film_actors;
```

This returns a sorted list of deduplicated actor names:

actors varchar[]
[Ana de Armas, Christoph Waltz, Daniel Craig, Léa Seydoux, Margot Robbie, Ryan Gosling]

Figure 6.11 – A sorted list containing every distinct actor name in our dataset

The ability to apply list function calls is a powerful technique that enables you to capture complex logic concisely. However, as you wrap increasing numbers of function calls, they can become harder to read, with increasingly nested parentheses that become harder to parse mentally. This is where another DuckDB SQL enhancement can help produce more elegant and maintainable SQL queries. DuckDB allows you to chain successive applications of scalar functions using the dot (.) operator. Here's what our previous query looks like when rewritten using DuckDB's function chaining:

```
SELECT list(actor_name).list_distinct().list_sort() AS actors
FROM film_actors;
```

Function chaining is a particularly convenient DuckDB SQL enhancement that we'll cover again in *Chapter 10*, where we'll discuss a range of techniques for effective DuckDB use. For an overview of the highlights of DuckDB's friendly SQL enhancements, refer to the *DuckDB features* section in *Chapter 1*.

We've only covered a handful of functions that DuckDB provides for working with lists. We encourage you to explore the DuckDB documentation for nested functions, which outlines all the available functions for working with lists, as well as functions for working with other DuckDB nested data types: `https://duckdb.org/docs/sql/functions/nested.html`.

Next, we'll cover another convenient feature of DuckDB's friendly SQL enhancements: list comprehensions.

List comprehensions

When working with lists, you may find yourself needing to apply custom transformations to each value of the list, as well as sometimes needing to filter out elements of lists. DuckDB provides a convenient list comprehension syntax, which allows you to do both of these things together. This is modeled off the list comprehension syntax found in Python. For example, given an `actor` column that contains a list of VARCHAR values, such as in our previous results, here's a list comprehension expression that converts the string values in each list into lowercase using DuckDB's `lower` function:

```
[lower(actor) FOR actor IN actors]
```

You can think of list comprehension as building a new list based on an existing list by applying a custom operation to each value in the original list. The expression that occurs to the left of the FOR keyword – `lower(actor)`, in this case – specifies the desired transformation of each element in the original list, which is specified by the expression at the end of the comprehension – `actors`, in this case. The variable name that's used to capture each element of the list is up to you to define. Here, we have used `actor` – you just need to be consistent in referring to this variable in the transformation expression on the left of the FOR keyword.

To see a concrete example of list comprehension in action, let's say we want to get the last component of each actor's name from our list, converted into lowercase. The sequence of operations we'll apply to each element in our actor names list to achieve this is as follows:

1. Split the actor-name VARCHAR value on space characters into a LIST data type of name components using DuckDB's `split` function.

2. Retrieve the last element of the resulting LIST data type using positional list-indexing. We do this by using an index value of -1.

3. Apply the `lower` function to the VARCHAR value that was returned from the previous step.

Let's put these steps together into a query that uses a list comprehension to perform this transformation on each actor's name. We'll base this query on our list aggregation query, which produces a single list containing every actor name in the dataset. This means the `list(actor_name)` expression will be the target for our list comprehension. We'll also use DuckDB function chaining to define our transformations on each element, which will help make our transformation steps readable. Here's our query:

```
SELECT
    [actor.split(' ')[-1].lower()
    FOR actor IN list(actor_name)
    ] AS actor_last_name
FROM film_actors;
```

When we run this query, we'll see our single row of all the actors in the dataset, but this time with their names transformed into lowercase forms of their last names:

actor_last_name varchar[]
[craig, armas, craig, seydoux, waltz, robbie, gosling]

Figure 6.12 – The list of actor names, transformed to contain the last component of each name, converted into lowercase

In addition to transforming the values inside lists, we can also use list comprehensions to filter out elements based on a specified constraint. For example, let's say we wanted to filter our list of actor names so that we only include actors whose full names are longer than 12 characters. We'll create a query that uses list comprehension to apply this filtering step, while also transforming the resulting elements so that they're in uppercase. Here's what this looks like:

```
SELECT
    [actor.upper()
    FOR actor IN list(character_name)
    IF length(actor) > 12
    ] AS long_characters
FROM film_actors;
```

As we can see, the conditional component of list comprehension takes the form of an expression that we place after the IF keyword at the end of the list comprehension expression. In this case, we've used the length function – which returns the number of characters in a string – to provide a constraint that each string in the input list must be greater than 12 characters so that it can be included in the transformed list that our list comprehension produces. It turns out that only one element of our list meets this criterion, so the result of our query is a single uppercased name:

long_characters varchar[]
[MADELEINE SWANN]

Figure 6.13 – The result of filtering the list of actors based on character length and converting them into uppercase

DuckDB's list comprehensions provide an elegant and concise way to apply custom transformation over a LIST data type. As we've seen, they become especially powerful when we make use of DuckDB's rich library of built-in functions, with DuckDB's function-chaining syntax providing a convenient way to compose multiple function applications. We focused on a handful of string manipulation functions here, but there are many more functions both for string manipulation and for working with DuckDB's other data types. You might like to explore the full catalog of DuckDB functions, broken down by working with different types of data: https://duckdb.org/docs/sql/functions/overview.

This brings us to the end of our exploration of working with lists in DuckDB. While brief, we hope it has given you a sense of the power and utility of DuckDB's LIST data type. Its flexibility, combined with DuckDB's rich selection of list functions and friendlier SQL enhancements, enables particularly effective modeling and processing of semi-structured data that involves variable-length sequences of values.

Working with JSON

JSON is a ubiquitous data interchange format that you will likely encounter in your adventures with analytical data, whether sourced from the result of API calls, extracted from other databases, or used as your target data-serialization format. As a human-readable text format with the flexibility of a self-describing schema, it is perhaps one of the most frequent sources of semi-structured data that you're likely to encounter. DuckDB provides a range of SQL functions via its json extension, which enables you to effectively read, transform, and write JSON data. In this section, we'll go on a short tour of using DuckDB to work with JSON data, highlighting some key features.

The JSON extension

DuckDB's json extension provides a range of SQL functions for working with JSON data. It also provides the JSON logical data type for representing and working with JSON values. Under the hood, DuckDB stores JSON values as strings; the JSON logical type allows you to work conveniently with JSON values, with DuckDB automatically performing JSON serialization and deserialization for you. The need to parse string values does mean that there is a performance hit to working with JSON values, so for complex querying or transformation-heavy workloads, you may want to consider preparing JSON data so that it's in a conformed tabular representation or converting your JSON data into a more performant nested data type, such as STRUCT or MAP, if this is supported by your data's schema.

In addition to the JSON data type, the json extension provides a rich collection of functions for working with JSON data. This includes functions for the following:

- Reading and writing JSON files

- Getting information about stored JSON values

- Extracting information from JSON values using JSON path queries

- JSON data type aggregation functions

- Serializing SQL statements to JSON values and deserializing them back to SQL

The json extension is bundled with almost every DuckDB client (including the DuckDB CLI) and is pre-loaded on DuckDB startup. This means that even though DuckDB's JSON features are made available through an extension, you will likely never need to explicitly install or load the extension. If you ever need to do this, refer to *Chapter 5*, where we covered installing and loading DuckDB extensions.

Creating JSON objects

We'll start our exploration of working with JSON data by looking at how to create JSON objects from existing tabular DuckDB data. We'll look at how we can do this using both scalar transformations – converting data row-by-row into corresponding JSON value rows – as well as through aggregation, where we'll group records and collapse each group into JSON values corresponding to the data in each group.

We'll use the film_actors table that we created previously, which contains a dataset of actor-name occurrences in films and the characters they played. Let's start by converting each row in this table into a single JSON object containing a subset of fields in the dataset. We can do this using the json_object function, which requires us to specify the key name and corresponding values for each property in the resulting JSON object. The function expects us to pass this as a sequence of key-value pairs.

Let's look at a query that creates JSON objects for each row in our film-actors dataset:

```
SELECT json_object(
    'film_name', film_name,
    'actor_name', actor_name
) AS json_created
FROM film_actors;
```

Note that we've provided string values for the key names, and appropriate columns as the corresponding values. This query returns a table with a single `film_actors` column with a data type of the JSON logical type:

json_created json
{"film_name":"James Bond - No Time to Die","actor_name":"Daniel Craig"} {"film_name":"James Bond - No Time to Die","actor_name":"Ana de Armas"} {"film_name":"James Bond - Spectre","actor_name":"Daniel Craig"} {"film_name":"James Bond - Spectre","actor_name":"Léa Seydoux"} {"film_name":"James Bond - Spectre","actor_name":"Christoph Waltz"} {"film_name":"Barbie","actor_name":"Margot Robbie"} {"film_name":"Barbie","actor_name":"Ryan Gosling"}

Figure 6.14 – Film and actor records as JSON objects

Having seen how we can use the `json_object` scalar function to create individual JSON records corresponding to rows, let's see how we can use a JSON aggregate function to produce JSON objects corresponding to groups of rows. Let's say we needed to produce a JSON object for each movie that contains each actor and their character as key-value pairs. We can use the `json_group_object` aggregate function, along with a GROUP BY function on the `film_name` column, to marshal a collection of desired attributes for each movie:

```
SELECT film_name,
    json_group_object(
        actor_name,
        character_name
    ) AS actor_character_json
FROM film_actors
GROUP BY film_name;
```

This query gives us a row for each movie, with our desired JSON object in each row:

film_name varchar	actor_character_json json
James Bond - Spectre Barbie James Bond - No Time to.	{"Daniel Craig":"James Bond","Léa Seydoux":"Madeleine Swann","Christoph Waltz":"Blofeld"} {"Margot Robbie":"Barbie","Ryan Gosling":"Ken"} {"Daniel Craig":"James Bond","Ana de Armas":"Paloma"}

Figure 6.15 – Actors and their roles, grouped by movie

Our query performed an aggregation over the original seven rows, combining the actor occurrences into three rows – one for each movie. The resulting `actor_character_json` column contains our desired JSON objects, which contain name-value attributes mapping the `actor_name` field to the `character_name` field.

It's worth remembering that JSON objects are generally slower to work with than a DuckDB STRUCT data type, which may look superficially similar. The consistency of the STRUCT object allows DuckDB to perform several optimizations, meaning they are almost always more efficient and faster to manipulate than JSON objects. For example, if we wanted to prepare a dataset containing semi-structured records of the same form as our previous examples that supports heavy query or transformation workloads, it would be more appropriate to use the more optimized STRUCT data type, given that we know each row will have the same schema.

As a brief aside, DuckDB allows us to readily convert a structured table of columns into rows of a STRUCT data type containing column values as key-value pairs. To do this, we simply need to SELECT the name of the target table, at which point DuckDB will convert each row into a single STRUCT value. Let's see this in action with our `film_actors` table:

```
SELECT film_actors
FROM film_actors;
```

This query yields the contents of our previous table, but with each row now taking the form of a single STRUCT value:

film_actors struct(film_name varchar, actor_name varchar, character_name varchar)
{'film_name': James Bond - No Time to Die, 'actor_name': Daniel Craig, 'character_name': James Bond}
{'film_name': James Bond - No Time to Die, 'actor_name': Ana de Armas, 'character_name': Paloma}
{'film_name': James Bond - Spectre, 'actor_name': Daniel Craig, 'character_name': James Bond}
{'film_name': James Bond - Spectre, 'actor_name': Léa Seydoux, 'character_name': Madeleine Swann}
{'film_name': James Bond - Spectre, 'actor_name': Christoph Waltz, 'character_name': Blofeld}
{'film_name': Barbie, 'actor_name': Margot Robbie, 'character_name': Barbie}
{'film_name': Barbie, 'actor_name': Ryan Gosling, 'character_name': Ken}

Figure 6.16 – Automatic STRUCT creation

While each row's single value looks very much like a JSON object, from the results, we can see that this column has a STRUCT type, with the field names and values being automatically inferred by DuckDB.

We haven't dived into working with STRUCT usage in as much detail as we have with the LIST and JSON objects, so we very much encourage you to explore DuckDB's documentation around working with STRUCT usage yourself: `https://duckdb.org/docs/sql/data_types/struct`. Next, we'll continue working with JSON data and focus on importing existing JSON data into DuckDB.

Importing JSON data

In *Chapter 2*, we briefly introduced loading data from JSON files. Here, we'll dive a bit deeper into a hands-on example of loading newline-delimited JSON data. We have been provided with the `media_tv.json` file, which contains information about a small collection of television shows. Each record is a complex data structure describing the style of the show, including the release date and scheduling information. An example of one of these records for the animated television show *The Simpsons* is presented here:

```json
{
  "media_type": "tv",
  "name": "The Simpsons",
  "media_payload": {
    "type": "Animation",
    "genres": [
      "Comedy",
      "Family"
    ],
    "premiered": "1989-12-16",
    "schedule": {
      "time": "19:00",
      "days": [
        "Sunday"
      ]
    }
  }
}
```

Figure 6.17 – A formatted JSON object containing a record of a television show

This is a detailed JSON record, so let's break it down so that we understand its components:

- `media_type`: A string indicating that the record is for a television show

- `name`: A string representing the name of the television show

- `media_payload`: An object containing a nested set of properties:

  - `type`: A string indicating the type of television show, such as `Animation`, `Scripted`, or `Live`

  - `genres`: A list of strings indicating the genres the show falls into, such as `Comedy`, `Family`, or `Romance`

  - `premiered`: The date of the first showing of the television show

  - `schedule`: An object that contains nested properties about the airing schedule, including the time of showing, and which days it's scheduled for

The media_tv.json file is formatted as newline-delimited JSON data, which means that each line of the file contains a complete JSON object for each record in the collection. By default, DuckDB's read_json function can automatically detect this type of formatting, so it turns out that loading these records into DuckDB is as simple as doing the following:

```
SELECT *
FROM read_json('media_tv.json');
```

This query returns a cleaned table of television records as rows, with appropriate column types inferred. Notice that the media_payload file has been converted into a STRUCT data type. This is an appropriate type for this complex property, which contains nested attributes:

media_type varchar	name varchar	media_payload struct("type" varchar, genres varchar[], premiered date, schedule struct("time" varchar.
tv	The Simpsons	{'type': Animation, 'genres': [Comedy, Family], 'premiered': 1989-12-16, 'schedule': {'.
tv	Bluey	{'type': Animation, 'genres': [Kids], 'premiered': 2018-10-04, 'schedule': {'time': 08:.
tv	Friends	{'type': Scripted, 'genres': [Comedy, Romance], 'premiered': 1994-09-20, 'schedule': {'.

Figure 6.18 – Television show records queried from a JSON file

This was a pleasantly concise query that enabled us to read and parse our television show JSON records into a tabular representation for working within DuckDB. There's a little bit of smarts that the read_json function uses under the hood to enable this convenience via the JSON parser's auto-detection features. Sometimes, you'll need to provide manual configuration to the JSON parser to get your desired results, so let's break down some of the configuration that's involved:

- By default, the autodetection feature of the JSON parser is turned on, which means that DuckDB will infer the names of the properties in your JSON data to use as columns, as well as their data types. It also enables the parser to detect the presence of newline-delimited data, interpreting each line as a distinct record. You can disable autodetection by passing auto_detect=false, but this also means that you'll need to supply an explicit schema for the JSON objects via the columns parameter.

- The parameter that you can use to instruct DuckDB to treat the target file as newline-delimited JSON objects is format. You can do this by passing it the 'newline_delimited' string value. We didn't need to do this in our example as autodetection inferred this value for us.

- In our results, you'll notice that read_json has returned a table that contains detected fields from the JSON records as columns, as opposed to simply mapping the JSON object to a single equivalent DuckDB native data type, which in this case would be STRUCT. This behavior is controlled by the records parameter, which is set to true by default. You can overwrite this if you just want to get a STRUCT data type back for each row by passing records=false.

It's worth spending some time familiarizing yourself with the effects of these parameters. For example, try seeing what happens when you set records=false in our previous query.

A situation you may find yourself in when reading JSON data is needing to override the automatically inferred schema by providing an explicit schema through the `columns` parameter. To illustrate how you can provide an explicit schema, let's modify our previous query by explicitly providing the schema definition that the autodetection feature inferred for us. Here's what this query looks like:

```
SELECT *
FROM read_json(
    'media_tv.json',
    columns={
        media_type: 'VARCHAR',
        name: 'VARCHAR',
        media_payload: 'STRUCT(
            type VARCHAR,
            genres VARCHAR[],
            premiered DATE,
            schedule STRUCT(
                time VARCHAR,
                days VARCHAR[]
            )
        )'
    }
);
```

A `STRUCT` data type is a natural fit for the nested `media_payload` field as we have a complex relationship to represent that involves show types, genres, and schedules. A lot is going on within this query, so let's highlight some key details:

- We have provided simple `VARCHAR` types for the `media_type` and `name` attributes.
- The `media_payload` column is specified as a `STRUCT` value that contains multiple properties that we must provide a schema for. They contain both field names and types.
- A `STRUCT` data type within the `media_payload` property is defined to store the properties of the `schedule` field, which is composed of `time` and `days` fields.
- Within the `media_payload` schema, we can see that the `genres` and `days` fields are of the `VARCHAR []` type, which indicates that its value is a `LIST` data type of `VARCHAR` values. This allows us to model shows as having multiple genres and multiple screening days.

Since we haven't changed this schema from the one that's automatically inferred in our initial query, the results for this will be identical. This should give you a sense of how you can customize the schema according to your needs. For example, you could change the target data types for specific values, or you could remove some of the properties from the schema and they would be omitted from the resulting query results. This also serves to illustrate some of the convenient heavy lifting that the autodetection feature of DuckDB's parser is doing, saving you from having to identify and type out the schema when you don't need to perform any customization.

Before we move on, let's store the result of our initial JSON loading query as a table called `media` in our database:

```
CREATE OR REPLACE TABLE media AS
SELECT *
FROM read_json('media_tv.json');
```

Using our media table, let's run a `DESCRIBE` statement to inspect its properties:

```
DESCRIBE media;
```

This gives us detailed information about the table, including the data types of each column:

column_name varchar	column_type varchar	null varchar	key varchar	default varchar	extra varchar
media_type	VARCHAR	YES			
name	VARCHAR	YES			
media_payload	STRUCT("type" VARCHAR, genres VARCHAR[], premiered DATE, sc.	YES			

Figure 6.19 – The result of describing our media table

Before we wrap up our coverage of importing JSON data, let's explore some queries for retrieving data from this table. Note that we now have a dataset that no longer contains any JSON typed data. It does, however, contain a `STRUCT` data type. Given that this will often occur when you're importing semi-structured data from JSON into DuckDB, it's worth spending a moment looking at how we can query from DuckDB tables involving a `STRUCT` data type.

Let's perform a simple query on our media table to retrieve its rows:

```
SELECT name, media_payload
FROM media;
```

Notice that this query returns the `media_payload` column as-is, giving us back the complete `STRUCT` for each row. While the representation of our `STRUCT` looks like `JSON`, the column type confirms that they are indeed stored as `STRUCT`:

name varchar	media_payload struct("type" varchar, genres varchar[], premiered date, schedule struct("time" varchar, "days" varchar[]))
The Simpsons	{'type': Animation, 'genres': [Comedy, Family], 'premiered': 1989-12-16, 'schedule': {'time': 19:00, 'days': [Sunday]}}
Bluey	{'type': Animation, 'genres': [Kids], 'premiered': 2018-10-04, 'schedule': {'time': 08:00, 'days': [Saturday, Sunday]}}
Friends	{'type': Scripted, 'genres': [Comedy, Romance], 'premiered': 1994-09-20, 'schedule': {'time': 20:00, 'days': [Thursday]}}

Figure 6.20 – Data loaded into the media table

If we wish to query or interact with the data within the `media_payload` field, we're going to need to extract parts of the `STRUCT` data type.

One way of extracting the components of `STRUCT` is to query the keys of the `media_payload` column using the dot (`.`) operator. For example, to retrieve the value of the `genres` key as a column in our results, we can `SELECT` on `media_payload.genres`. While applying this pattern to extract all values of each `STRUCT` data type, we can do the following:

```
SELECT name,
    media_payload.type,
    media_payload.genres,
    media_payload.premiered,
    media_payload.schedule
FROM media;
```

By selecting each field in `media_payload`, we've converted the nested values in this `STRUCT` data type into distinct columns in our results:

name varchar	type varchar	genres varchar[]	premiered date	schedule struct("time" varchar, "days" varchar[])
The Simpsons	Animation	[Comedy, Family]	1989-12-16	{'time': 19:00, 'days': [Sunday]}
Bluey	Animation	[Kids]	2018-10-04	{'time': 08:00, 'days': [Saturday, Sunday]}
Friends	Scripted	[Comedy, Romance]	1994-09-20	{'time': 20:00, 'days': [Thursday]}

Figure 6.21 – The result of querying individual keys from the media_payload column

If you felt that this query was a bit verbose, DuckDB comes to the rescue here with some convenient syntax. We can use the `*` shortcut to unpack a `STRUCT` data type into separate columns, without having to explicitly name each key. This is a particularly helpful shortcut for dynamically selecting all keys present within `STRUCT` objects. Using this trick, we can simplify the earlier query, and avoid the need to list each key field, by selecting the `media_payload.*` expression:

```
SELECT name, media_payload.*
FROM media;
```

This query gives us the same result as before but with a much simpler query.

The `*` shortcut for unpacking a `STRUCT` data type is a great time saver when it comes to extracting the fields of a complex or perhaps unknown schema, and it's worth keeping in your bag of DuckDB tricks.

Next, we'll continue this tangent of working with semi-structured data that we've generated from imported JSON data. In particular, we'll look at how DuckDB allows us to unpack nested values within `LIST` and `STRUCT` data types into multiple rows.

Unpacking lists with the unnest function

Earlier, we learned how to aggregate rows into lists using the `list` aggregation function. There, we nested multiple rows into a LIST data type. DuckDB also allows us to do the reverse – transform the nested data types of LIST or STRUCT and expand them into rows through the unnest function. Like aggregate functions, unnest changes the cardinality of our results, but rather than reducing a group of rows into a single row, the unnest function expands a single row containing a nested value, into multiple rows, each containing a value from the nested data structure.

Let's demonstrate the unnest function by creating a table called `media_extracted` that contains television show metadata:

```
CREATE OR REPLACE TABLE media_extracted AS
SELECT name, media_payload.*
FROM media;
```

With the `media_extracted` table created, we can review the available attributes with a query:

```
SELECT *
FROM media_extracted;
```

We'll get the following result:

name varchar	type varchar	genres varchar[]	premiered date	schedule struct("time" varchar, "days" varchar[])
The Simpsons	Animation	[Comedy, Family]	1989-12-16	{'time': 19:00, 'days': [Sunday]}
Bluey	Animation	[Kids]	2018-10-04	{'time': 08:00, 'days': [Saturday, Sunday]}
Friends	Scripted	[Comedy, Romance]	1994-09-20	{'time': 20:00, 'days': [Thursday]}

Figure 6.22 – The contents of our media_extracted table

As we can see, the `media_extracted` table has a `genres` column that contains a LIST data type of genre names as VARCHAR values. For example, *The Simpsons* can be described as being part of both the *Comedy* and *Family* genres.

Let's use the unnest function to unpack the values in each list of genres. This will expand our number of rows by creating a record for each show/genre pair that occurs in the dataset:

```
SELECT name, unnest(genres)
FROM media_extracted;
```

This query returns five rows, with each row representing a show name and genre-type pair. This is two more rows than we had in our starting table, and they correspond to the two shows that have two listed genres:

name varchar	unnest(genres) varchar
The Simpsons	Comedy
The Simpsons	Family
Bluey	Kids
Friends	Comedy
Friends	Romance

Figure 6.23 – Expanded rows resulting from unnesting the show genres

This can be a useful technique for processing semi-structured information, but you need to ensure the change in cardinality (number of rows) makes logical sense in the context of your query. It should also be noted that using unnest does not preserve the order of the list – so the rows will appear in an arbitrary order unless an ORDER BY clause is used in the SQL statement.

Having now looked at some examples of working with DuckDB-native semi-structured data that we've imported from JSON data, let's return to working directly with JSON data and the JSON logical type by seeing how DuckDB's json extension allows us to work effectively with JSON records containing inconsistent schemas.

Working with inconsistent schemas

A common complication when working with semi-structured data is encountering datasets containing records that follow multiple schemas. A common scenario is having to work with a collection of JSON data containing records that have been collected from various sources. In this situation, you will likely have to manage the fact that records in this dataset will often conform to different schemas, making their normalization challenging.

Let's build a scenario for DuckDB where we wish to load and parse inconsistent JSON schemas representing television and movie data sourced from a single file containing divergent records. We'll start by inspecting the media_mixed.json file, which contains a variety of data, including records corresponding to both films and television shows. In this small dataset, both types of records have a media_type property, indicating whether they are film or TV, and both have a media_payload property, which contains information about the piece of media. The schema for this field varies according to the value of the media_type field. Reading the contents of this file via a simple SELECT statement reveals the divergent nature of these two schemas:

```
SELECT media_type, media_payload
FROM read_json('media_mixed.json');
```

When the `media_payload` field is converted into a JSON object for each record, we can see the divergent schemas clearly, with missing fields being included as explicit NULL values:

media_type varchar	name varchar	media_payload struct(first_film_screened date, staring varchar[], "type" varchar, genres varc.
film	James Bond - No Ti.	{'first_film_screened': 2022-10-05, 'staring': [Daniel Craig, Ana de Armas], 't.
film	Barbie	{'first_film_screened': 2023-02-05, 'staring': [Margot Robbie, Ryan Gosling], '.
tv	The Simpsons	{'first_film_screened': NULL, 'staring': NULL, 'type': Animation, 'genres': [Co.
tv	Bluey	{'first_film_screened': NULL, 'staring': NULL, 'type': Animation, 'genres': [Ki.
tv	Friends	{'first_film_screened': NULL, 'staring': NULL, 'type': Scripted, 'genres': [Com.

Figure 6.24 – Query results showing the contents of the media_payload object as a STRUCT data type

To make the differences between the two schemas clearer, let's examine a film record and a TV record side by side. Notice the difference in the contents of the `media_payload` object across the two records:

film	tv

``
{
  "media_type": "film",
  "name": "James Bond – No Time to Die",
  "media_payload": {
    "first_film_screened": "2022-10-05",
    "staring": [
      "Daniel Craig",
      "Ana de Armas"
    ]
  }
}
``
	``
	{
	  "media_type": "tv",
	  "name": "Bluey",
	  "media_payload": {
	    "type": "Animation",
	    "genres": [
	      "Kids"
	    ],
	    "premiered": "2018-10-04",
	    "schedule": {
	      "time": "08:00",
	      "days": [
	        "Saturday",
	        "Sunday"
	      ]
	    }
	  }
	}
	``

Figure 6.25 – Comparing film and television show records

Visually inspecting differences in schemas is feasible for a handful of records, but what if we were tackling a much larger dataset that could potentially have even more diverging schemas? The json extension gives us a way to inspect the structure of JSON objects programmatically through the json_group_structure aggregate function. This function allows us to readily compare the structure of JSON objects across different groupings in our dataset. Here, we want to compare the structure found in the media_payload object across the different media types, so we will use media_payload as the argument to the json_group_structure function and use GROUP BY on media_type. Here's our query:

```
SELECT media_type, json_group_structure(media_payload)
FROM read_json('media_mixed.json')
GROUP BY media_type;
```

This query gives us a clear indication of the schemas across both types of records, including where they diverge, through the appearance of NULL values:

media_type varchar	json_group_structure(media_payload) json
film	{"first_film_screened":"VARCHAR","staring":["VARCHAR"],"type":"NULL","genres":"NULL","premiered":"NULL.
tv	{"first_film_screened":"NULL","staring":"NULL","type":"VARCHAR","genres":["VARCHAR"],"premiered":"VARC.

Figure 6.26 – Programmatically extracted schemas, grouped by media type

The preview of the results of this query has been truncated in *Figure 6.27* as there's a fair bit of detail. You may wish to inspect this in more detail for yourself. The key thing to observe is that if we had a larger number of media types with diverging schemas, this technique would provide a much more effective way to understand differences across schemas within a collection of heterogeneous JSON objects.

Now that we have a grasp on the structure of our inconsistent schemas, we can readily extract cleaned records corresponding to the different types of records. To extract film records, we can simply query the media_mixed.json file for JSON objects whose media_type is film, while only selecting fields these records contain:

```
SELECT media_type,
    name,
    media_payload.first_film_screened,
    media_payload.staring
FROM read_json('media_mixed.json')
WHERE media_type = 'film';
```

This produces a nice clean table of film rows containing only columns about film records that DuckDB has automatically detected appropriate types for.

Running this query gives us clean, structured results for film records and their attributes:

media_type varchar	name varchar	first_film_screened date	staring varchar[]
film film	James Bond - No Time . Barbie	2022-10-05 2023-02-05	[Daniel Craig, Ana de Armas] [Margot Robbie, Ryan Gosling]

Figure 6.27 – Cleaned film records

Let's do the same again, but this time for records with `media_type` set to `tv`:

```
SELECT media_type,
    name,
    media_payload.type,
    media_payload.genres,
    media_payload.premiered,
    media_payload.schedule,
FROM read_json('media_mixed.json')
WHERE media_type = 'tv';
```

Again, we can see that we've produced a cleaned set of rows, this time corresponding to television show records that only contain columns appropriate for this media type:

media_type varchar	name varchar	type varchar	genres varchar[]	premiered date	schedule struct("time" varchar, "days" varchar[])
tv	The Simpsons	Animation	[Comedy, Family]	1989-12-16	{'time': 19:00, 'days': [Sunday]}
tv	Bluey	Animation	[Kids]	2018-10-04	{'time': 08:00, 'days': [Saturday, Sunday]}
tv	Friends	Scripted	[Comedy, Romance]	1994-09-20	{'time': 20:00, 'days': [Thursday]}

Figure 6.28 – Cleaned television show records

From these results, you might have noticed that we didn't unpack the `schedule` object, so DuckDB has left this as a semi-structured value by converting it into a `STRUCT` data type. You might like to try extending this example by fully unpacking the nested properties of the `schedule` property into separate columns.

The key thing to take away from this set of exercises is that DuckDB's `JSON` logical type, along with its supporting JSON functions, provides us with a convenient way to analyze and then reach into a collection of heterogeneous JSON records and extract conformed and well-typed tables. This challenge of working with inconsistent schemas is one that often arises when working with JSON data collected from API services. In the next section, we'll look at how we can use DuckDB to query a public API and then work with the JSON data it returns.

Querying JSON data from an API

Before we end this chapter, we'll cover one more topic: querying JSON data retrieved from a remote API. We can use DuckDB's ability to work with data returned by a programming interface.

In *Chapter 5*, we encountered DuckDB's `httpfs` extension, which enables us to read remotely hosted files and interact with object storage using the S3 API. One of the applications of this functionality is retrieving and querying JSON data exposed by an HTTP API.

As an example of a publicly available REST API that's exposed over HTTP, we'll be using TVmaze, a free user-curated television database service. Their API (`https://www.tvmaze.com/api`) is a free service that allows us to query for information about television shows, episodes, actors, and scheduling information. As TVmaze's API returns JSON data, we can query the API using DuckDB via the `httpfs` extensions, and then work with the results using the `json` extension.

Let's use the TVmaze API to discover some details about the television show *The Simpsons*. We'll initially be calling the `singlesearch` endpoint while searching for the `The%20Simpsons` query string (here, `%20` represents the space character in an encoded URL):

```
SELECT *
FROM read_json(
     'https://api.tvmaze.com/singlesearch/shows?q=The%20Simpsons');
```

This query may take a moment to complete as it is interacting with a remotely hosted API, but you should retrieve a response like this:

id int64	url varchar	name varchar	…	image struct(medium varc…	summary varchar	updated int64	_links struct(self struct…
83	https://www.tvmaze…	The Simpsons	…	{'medium': https:/…	\<p>\<b>The Simpsons…	1694755583	{'self': {'href': …

Figure 6.29 – Show information returned from the TVmaze API

As we can see, DuckDB has retrieved the JSON response and has inferred a schema from the API. In the response, the TVmaze API has provided us with the show ID of `83` for *The Simpsons*, which we can use to perform further queries for this show.

Let's retrieve more information about this show from a different TVmaze endpoint. We'll construct a query that fetches information about show details for season 34, episode two of *The Simpsons*. Specifically, we'll extract the average audience rating for this episode, which is included in the JSON payload response for the endpoint we'll query.

The json_extract_string function allows us to provide a JSON value and **JSONPath** expression, which specifies the element we want to extract from the JSON value. The path expression must be provided as a string, which follows JSONPath syntax. This is a small domain-specific language that supports concise queries to extract elements within nested JSON objects. In this case, we want to extract the average user rating for the episode, which is contained in a nested average property, inside the rating property of the JSON response. To extract this value, we'll pass the rating property as the first argument to the json_extract_string function, and the $.average JSONPath expression, which indicates we wish to extract the average field from inside the rating property, with the $ symbol to indicate the root element. Here's our complete query:

```
SELECT season,
    number,
    name,
    json_extract_string(rating, '$.average') AS avg_rating,
    summary
FROM read_json('https://api.tvmaze.com/shows/83/
episodebynumber?season=34&number=2');
```

After a moment, the query will return a row of information for our target episode. Note that your results may be different as the average rating may have changed over time:

season int64	number int64	name varchar	avg_rating varchar	summary varchar
34	2	One Angry Lisa	6.5	\<p>Lisa gets called for jury duty while Marge becomes obsessed with her exercise bi...

Figure 6.30 – Episode information returned from the TVMaze API

Again, DuckDB has retrieved the JSON response and has inferred a schema from the JSON payload that was returned from the API. On top of this, we also used the json_extract_string function to extract embedded audience feedback on the episode.

DuckDB's ability to work with data returned by a programming interface is rather useful, allowing us to explore data beyond file and table-based structures. The ability to query JSON from a remote API coupled with DuckDB's semi-structured data processing functions enables us to analyze a variety of rich and complex data services.

Summary

In this chapter, we saw how DuckDB offers a range of data types, functions, and SQL enhancements that make it a versatile data processing tool for working with semi-structured data. We covered some of DuckDB's key nested data types – LIST, MAP, and STRUCT – which we then built on so that we could explore working with semi-structured data. This included working with DuckDB's LIST data type and seeing the flexibility it offers, as well as covering a selection of list processing functions and SQL enhancements that enable some powerful and effective list processing patterns. We then turned to look at DuckDB's json extension and learned how we can create JSON objects from existing DuckDB data, import JSON data into DuckDB, and work with semi-structured data produced from JSON imports. We also saw how DuckDB enables us to work effectively with JSON data containing inconsistent schemas. We concluded this chapter with a set of practical examples that illustrated how we can use DuckDB to query remote APIs and work with the resulting JSON data.

This concludes our treatment of working with semi-structured data with DuckDB. In the next chapter, we'll switch gears a little, turning our attention to getting started working with the DuckDB Python client.

7

Setting up the DuckDB Python Client

Up until now, we have been working with DuckDB using its terminal-based CLI client. While the CLI client offers a convenient way to jump in and start using DuckDB, it does not lend itself as well to being integrated with environments and workflows for data analysis and for the development of operational data services or products. Python and its rich ecosystem of data packages and tooling is a popular programming language for both these families of use cases. The DuckDB Python client offers a feature-rich and versatile collection of APIs for interacting with DuckDB databases in Python. It also offers rich integrations with Python and its ecosystem, such as enabling you to register Python functions to be used in DuckDB queries and allowing you to consume from and export to in-memory data structures, including pandas dataframes, Polars dataframes, and Apache Arrow Tables.

We'll cover all these powerful features just mentioned in *Chapter 8*. In this chapter, we'll focus on the first stages of getting started using DuckDB with Python:

- Setting up your environment to work with DuckDB using Python
- Understanding the different ways to connect to DuckDB databases

By the end of this chapter, you'll be all set up with a working Python environment in which you can readily connect to and create new DuckDB databases. You'll then be ready to jump into the next chapter, where we'll continue our treatment of the DuckDB Python client, working with hands-on examples across different parts of DuckDB's Python APIs.

Technical requirements

As we're going beyond using the simple DuckDB CLI client, this chapter and the next have a bit more in the way of technical requirements than we've needed previously. This section will take you through the steps for getting an environment set up on your local machine, which you'll need to work through the examples in this chapter. You'll also need this for *Chapter 8*, where we'll dive into using DuckDB with Python, as well as *Chapter 11*, where we'll do some hands-on data analysis using DuckDB. We'll go into some detail when describing these steps, however, you may also want to refer to some of the more comprehensive resources that we provide pointers to.

The requirements you'll need to have set up, which we'll go through shortly, are as follows:

- A cloned copy of the GitHub repository for this book
- A Python installation (version 3.7+)
- A Python virtual environment with the required dependencies installed
- An **integrated development environment** (**IDE**) with support for Jupyter Notebooks

The examples for this chapter can be found in the `chapter_07` directory of this book's GitHub repository: `https://github.com/PacktPublishing/Getting-Started-with-DuckDB/tree/main/chapter_07`.

Cloning the repository

If you haven't already done so, you'll need to clone the repository for this book, which can be found at `https://github.com/PacktPublishing/Getting-Started-with-DuckDB`. You'll need to install Git if you don't have it on your machine yet (`https://git-scm.com/downloads`). See the GitHub documentation for detailed instructions on how to clone repositories from GitHub onto your machine: `https://docs.github.com/en/repositories/creating-and-managing-repositories/cloning-a-repository`.

Installing Python

As we'll be working with Python, you'll need Python installed on your machine. At the time of writing, DuckDB supports Python 3.7 and above. If needed, there are several ways to get a local Python installation on your machine. We suggest visiting `https://www.python.org/downloads` and then downloading and installing the latest version of Python for your operating system. Detailed guides for installing and configuring Python installations on Windows, macOS, and Linux systems can be found at `https://docs.python.org/3/using`. Alternatively, if you prefer to work with conda environments, you may want to install the Anaconda distribution of Python (`https://www.anaconda.com/download`). Note that the rest of these instructions will assume you're working with a standard Python installation (i.e., not Anaconda), so we will be using `pip` rather than `conda` to install dependencies.

Creating a Python virtual environment

If you have not already made one, we also recommend you create a Python virtual environment for this project. This will allow you to install the required dependencies you'll need for this chapter into a sandboxed environment that won't interfere with any installed dependencies you may need for other projects. For those new to working with Python virtual environments, we'll provide a short how-to. For a more comprehensive guide, see the Python documentation for the `venv` module: `https://docs.python.org/3/library/venv.html`.

To get your virtual environment set up, open your terminal application and change directory to the top-level directory of the `Getting Started with DuckDB` repository. We'll then create a virtual environment with the following command:

```
$ python -m venv duckdb-book
```

This will result in a virtual environment being created in a new directory named `duckdb-book`. In order to use this virtual environment, you then need to activate it using the appropriate command for your environment.

Activating on macOS or Linux is done as follows:

```
$ source duckdb-book/bin/activate
```

Activating on Windows using `cmd.exe` requires the following command:

```
C:\> duckdb-book\Scripts\activate.bat
```

When using Windows PowerShell, you may also need to run an additional command before you can run the activation script:

```
PS C:> Set-ExecutionPolicy -ExecutionPolicy RemoteSigned -Scope
CurrentUser
PS C:\> duckdb-book\Scripts\Activate.ps1
```

If you run into any issues with this, we suggest you consult the helpful `venv` module documentation. Once activated, you'll see an indication of the current virtual environment in parentheses to the left of the prompt in your terminal.

```
ned@localhost:Getting-started-with-DuckDB $ source duckdb-book/bin/activate
(duckdb-book) ned@localhost:Getting-started-with-DuckDB $
```

Figure 7.1 – Activating a Python virtual environment in a terminal on a Linux operating system

While active, any dependencies you install with `pip` will be installed only into this sandboxed environment, and Python processes launched within this virtual environment will be able to import from these dependencies. Note that you'll need to re-activate the virtual environment if you close your terminal. When you're finished using the virtual environment, you can simply exit the terminal, or if you want to keep using the terminal session for other activities, you can deactivate the virtual environment by running the following command:

```
$ deactivate
```

If you ran the `deactivate` command just now, don't forget to re-run the activation command before working through the rest of the chapter!

Installing the Python dependencies

With your `duckdb-book` virtual environment activated, the next step is to install the Python dependencies we'll need. We'll use pip, Python's package installer, to install the packages defined in the `requirements.txt` file from the repository. Make sure your active directory on the command line is the root directory of the cloned Git repository and run this command:

```
$ pip install -r requirements.txt
```

You can verify that the installation has worked by opening a Python session and checking that you can import DuckDB.

```
(duckdb-book) ned@localhost:Getting-started-with-DuckDB $ python

Python 3.11.4 (main, Jun 25 2023, 16:39:25) [GCC 12.2.0] on linux
Type "help", "copyright", "credits" or "license" for more information.
>>> import duckdb
>>>
```

Figure 7.2 – Terminal output showing opening a Python session and importing the duckdb module

For your reference, when you don't have a predefined `requirements.txt` file available like we do here, the DuckDB Python client can be installed on its own using the following command:

```
$ pip install duckdb
```

On running this command, pip will download the latest version of DuckDB from the Python Package Index, also known as PyPI (https://pypi.org). You may also wish to consult the DuckDB documentation for installing DuckDB clients: https://duckdb.org/docs/installation.

Choosing your Python IDE

Most of the exercises from this chapter can be run inside any interactive Python shell, however the bare-bones Python **read-eval-print loop** (**REPL**) can be challenging to use for interactive data-analysis workflows. When exploring a new dataset or performing an analysis of a dataset, a common approach is to develop a sequence of steps that build on each other, interactively inspecting their output as you proceed. Simple REPLs, however, typically don't have the rendering capabilities required for displaying rich data visualizations—a crucial component of data analysis. Also, REPLs are designed around a linear execution mode, which makes it tricky to re-run previous steps. When performing a data analysis, revisiting earlier steps to diagnose an issue or make modifications is a common occurrence.

For these reasons, we recommend using one of the various Jupyter Notebook-based IDEs, which are well suited for interactive non-linear workflows and offer excellent support for rendering rich and interactive data visualizations. This is what we'll be using for working on the Python examples throughout this book. If you don't already have a preferred IDE for working with Jupyter Notebooks, we suggest you use one of the following:

- JupyterLab

- Microsoft **Visual Studio Code** (**VS Code**)

JupyterLab is a fully-featured analytical IDE designed specifically for working with Jupyter Notebooks. If you are unsure which notebook IDE to use, we recommend using this. JupyterLab is included in `requirements.txt`, so you will already have this installed in your virtual environment. To start JupyterLab, simply run this command in your terminal:

```
$ jupyter-lab
```

This will start a Jupyter server in your terminal that you can connect to using your browser. If a browser window does not automatically pop up with JupyterLab running, the output of the server in your terminal will show you the URL of the local server, which you can copy and paste into your browser. While we believe JupyterLab offers a superior experience, some people may prefer the classic Jupyter Notebook experience rather than JupyterLab. You can launch Jupyter Notebook by running `jupyter-notebook` in your terminal instead.

Alternatively, if you already use and are familiar with VS Code, you may find that you prefer to use its Jupyter Notebook support. You can find instructions for how to set this up in the VS Code documentation: `https://code.visualstudio.com/docs/datascience/jupyter-notebooks`. Note that VS Code's Jupyter Notebook support still requires a Jupyter installation, which can be sourced from the `duckdb-book` virtual environment.

Creating a new notebook

After getting your environment and IDE set up, the last thing to do is create a new notebook in your chosen IDE. This is where you'll be entering and running the code that we'll be working through for both this chapter and the next. With all this set up, we're now ready to connect to a DuckDB database!

Connecting to DuckDB in Python

When working with DuckDB in Python, the first thing to consider is how you'll connect to a target DuckDB database from your Python code. Even though DuckDB runs embedded within a parent process, meaning that you don't have to worry about authenticating to a remote server, you still need to consider how your Python process will make and manage connections to potentially multiple DuckDB databases.

In this section, we'll be covering the two broad ways to issue commands to DuckDB databases in Python:

1. **Using the default in-memory database**: This database is automatically created for you when you import the `duckdb` module. The default database provides a convenient way to quickly perform activities such as ad-hoc data analysis when you know you will only need a single database in your session. This is what you will connect to when you invoke a function from the `duckdb` module that interacts with a database, such as `duckdb.sql()` and `duckdb.execute()`.

2. **Explicitly creating and managing database connections**: This gives you more flexibility and control over how you work with DuckDB databases. Database connection objects are created via the `duckdb.connect()` function, which you then use to issue commands against the database it's connected to.

Both approaches give you access to the same Python APIs for interacting with DuckDB databases, but with different levels of convenience and customizability. Let's go through each of these ways of interacting with DuckDB databases in Python.

Using the default in-memory database

The simplest way to hit the ground running using DuckDB in Python is to use the default database. This is a shared in-memory database that is available anywhere in your Python process after importing the `duckdb` module. It has the advantage of convenience, as it saves you from having to manage database connections yourself, making it perfect for ad-hoc data wrangling tasks, or if you're just taking DuckDB for a spin.

The DuckDB Python client provides a range of functions for interacting with DuckDB databases. When called as attributes of the `duckdb` module, they will dispatch your instructions to the shared in-memory database. From here on, we will refer to these functions associated with underlying connection objects as *methods*, in keeping with object-oriented programming terminology.

To see this in action, let's invoke the `duckdb.sql()` method. The `sql()` method is part of DuckDB's Relational API and is used to construct a SQL query for running against the database. The following code is a Python snippet that you will need to enter into and run inside a code cell of your Jupyter Notebook-compatible IDE. This is also how you should run all of the following code examples in this chapter:

```
import duckdb
duckdb.sql("SELECT 'duck' AS animal, 'quack!' AS greeting")
```

In most notebook-based IDEs, when you run a code cell, the value of the final expression is displayed in the cell's output area. We'll be leveraging this convenient feature throughout our notebook-based exercises to display the output of each example. This means that after you run this block, you'll see the following output:

| animal | greeting |
varchar	varchar
duck	quack!

Figure 7.3 – The result of a query run against the default database using the duckdb.sql() method

With that, we have just run our first DuckDB query in Python using the DuckDB Python client's default database. If you are wondering about the two different types of quote characters in the code we just ran, remember that in DuckDB SQL, literal string values are represented using single-quote characters. The outer double-quote characters define the Python string literal containing our SQL query that we want to send to the database. As Python can use either double quotes or single quotes to represent string literals, we recommend using double quotes for Python string literals that contain SQL queries, as this will save you from having to escape single quotes when using string literals in your SQL queries.

The result of the `sql()` method call is a DuckDB **relation** object, which represents our query. We'll unpack these relation objects in *Chapter 8*, as part of our treatment of DuckDB's Relational API. For now, we're just focusing on the fact that since we've invoked the `sql()` method against the `duckdb` module, our SQL query was automatically dispatched to the default database.

The default in-memory database is particularly convenient for interactive data analysis workflows, as you can jump in and start working with DuckDB without having to set up any connections. If you're doing ad-hoc data wrangling or exploratory data analysis in a notebook, you may often find yourself reaching for this approach. In other scenarios, the default database won't always be the most appropriate way to work with DuckDB. Being an in-memory database, the default database is ephemeral, meaning that its contents will be lost when its parent Python process ends. If you want to persist your database state after the process ends, you'll need to create a new persistent connection explicitly, rather than use the default database.

Another consideration is that the default database is shared globally across the duckdb module for each Python process. If you're developing a Python library that uses DuckDB, it's important to bear in mind that downstream users of your library may themselves want to use DuckDB's default database for their own needs. It is therefore recommended to avoid the default in-memory database when developing library packages, so that consumers of your library can safely import and use the duckdb module without clashing with any database state created by your library's code.

Next, we'll cover a more flexible way to connect to DuckDB databases that involves creating and configuring new connections explicitly. This approach allows you to manage and use multiple DuckDB databases in the same Python process, including persistent file-based databases, as well as allowing you to customize each databases' configuration.

Creating database connections explicitly

In addition to using the default in-memory database, you also have the option to use the duckdb. connect() function to explicitly create and manage connection objects. This is part of DuckDB's support of the Python DB-API 2.0 standard, which we'll cover in the next chapter. Managing database connection objects yourself gives you much more control over how you create and connect to the DuckDB database within your analytical environment or application logic. This is also how library packages can connect to DuckDB databases safely, without fear of clashing with any usage of DuckDB's default database that downstream users of the library might want to perform.

The connect() function returns a DuckDBPyConnection object that represents a connection to a specific DuckDB database. Calling connect() without any arguments results in the creation of a new in-memory database. It's important to note that this is distinct from the default in-memory database. Let's see this in action:

```
conn = duckdb.connect()
conn.sql(
    """
    CREATE TABLE hello AS
    SELECT 'pato' AS animal, 'cuac!' AS greeting
    """
)
conn.sql("SELECT * FROM hello")
```

This gives us the following output:

animal varchar	greeting varchar
pato	cuac!

Figure 7.4 – The result of a query run against a new in-memory
database, using the sql() method of a connection object

In this example, we created a new in-memory database and assigned the resulting connection object to the conn variable. We then submitted two SQL queries to this new database via the sql () method of the connection object: one that created a table with a single row and another that retrieved this single row.

The connect () function has several parameters, the first of which is the database parameter, which takes a string specifying the database you want to connect to. As we'll see later, passing a file path for this value results in a connection to a persistent database stored on disk. Its default value however is the special string :memory:, which is why calling the connect () function with no arguments returned us a connection to a new in-memory database. Let's make another in-memory database, specifying this explicitly with the database keyword argument:

```
another_conn = duckdb.connect(database=":memory:")
```

Running this code will result in a second in-memory database being created, distinct from the first one, and which we now have a reference to via the another_conn variable. Once again, this is distinct from the default in-memory database used by the duckdb.sql () method and other methods attached to the duckdb module. If you happen to want to create an explicit connection object to the default in-memory database, you can pass the :default: string as the value of the database keyword argument:

```
default_conn = duckdb.connect(database=":default:")
```

> **Every database interaction requires a connection**
>
> If you're finding the appearance of the `sql()` method on both the `duckdb` module and on connection objects confusing, it may be helpful to consider that every interaction with a DuckDB database using the Python client occurs via a `DuckDBPyConnection` object, even when issuing commands via methods from the `duckdb` module. When you call `duckdb.sql()`, it dispatches your query to the default database via a connection object that is automatically created when you import the `duckdb` module. This connection object is available via the `duckdb.default_connection` module attribute. To illustrate this, the following two Python expressions are functionally equivalent (but not in convenience):
>
> `duckdb.sql("SELECT 'hello'")`
>
> `duckdb.default_connection.sql("SELECT 'hello'")`
>
> Connection objects provide a range of methods beyond the `sql()` method for interacting with their database, all of which can be accessed via the `duckdb` module for convenient interaction with the default database. We'll see more of these methods in the next chapter.

Explicitly creating connection objects also allows us to specify database configuration options. To do this, you need to pass a dictionary of configuration key-value pairs to the `connect()` function's `config` parameter. Here's an example where we create a new in-memory database, which we configure to use a maximum of 10 GB of system memory, as well as limiting the database to using four CPU threads for parallel query execution:

```
custom_conn = duckdb.connect(
    config={
        "memory_limit": "10GB",
        "threads": 4
    }
)
```

These database configuration options may come in handy if you need to tune DuckDB's internal configuration to match the constraints of your environment. For best performance, the DuckDB performance guide recommends having 5 GB of memory available per thread for aggregation-heavy workloads, and 10 GB of memory per thread for join-heavy workloads, which is one reason why you might want to limit the number of threads DuckDB uses. Limiting the amount of system memory used by your database is perhaps a more niche need but may be helpful when working in multi-tenant environments with resource contention. Note that we have included these two configuration examples together in this example for illustration purposes, rather than targeting a specific use-case. See the DuckDB performance guide more information around tuning DuckDB's performance: `https://duckdb.org/docs/guides/performance`.

For the complete list of database configuration options that can be provided via the `config` parameter of the `connect` function, consult the DuckDB configuration documentation: `https://duckdb.org/docs/configuration`.

Connecting to persistent-storage databases

When using an in-memory database, no data is persisted to disk, and all data stored in the database is lost when the Python process exits. If you need your database contents to persist, you can create a disk-based DuckDB database by passing a file path to the connect() function's database parameter. If the database file already exists, a connection to this database will be created; if not, DuckDB will first create a new database at the specified path, before returning a connection to it. Let's look at an example, where we'll connect to a new on-disk database and write our simple table to it:

```
conn = duckdb.connect(database="quack.duckdb")
conn.sql(
    """
    CREATE OR REPLACE TABLE hello AS
    SELECT 'ente' AS animal, 'quak!' AS greeting
    """
)
conn.close()
```

After running this, you will now have a DuckDB database in the quack.duckdb file, which contains our hello table with a single row. Let's verify this quickly by running a shell command that invokes the DuckDB CLI to connect to the new quack.duckdb database and query the hello table. Rather than leaving the notebook, we'll do this in a notebook cell by prefixing our command with the ! character, which instructs the notebook to run it in a new shell, as if you had run it on your command line via your terminal application:

```
! duckdb quack.duckdb -c "SELECT * FROM hello"
```

Note that if you downloaded the DuckDB CLI directory, you'll need to replace duckdb with a relative path from the chapter_08 folder to the location of your DuckDB CLI executable. After running the command, we should see our new greeting:

animal varchar	greeting varchar
ente	quak!

Figure 7.5 – The result of a query run against a disk-based DuckDB database via the DuckDB CLI

This shows us that our database changes were successfully persisted in the DuckDB database file quack.duckdb.

Thinking about what we just did, notice that we connected to our on-disk database from a different process than the notebook's Python kernel. It's worth noting here that DuckDB does not allow other processes to read from a disk-based database that already has an existing write-enabled connection. This is why we made sure to run the `close()` method on our connection before querying it from another process. Without this step, our attempt to query the database using the DuckDB CLI would have thrown an error. In the next section, we'll discuss closing connections in more detail.

Closing database connections

It's good practice to make sure your connection objects are closed after you're finished working with them. This is particularly important when making changes to disk-based databases due to the fact that when writing to a database, DuckDB maintains a temporary data structure that needs to be synchronized with the database file to ensure data integrity. Ending your Python process without first closing any connections to database files could result in unsynchronized data failing to be written to the persistent database.

As we've already seen, the simplest way to close a connection is to call its `close()` method after you're finished using it:

```
sql = "INSERT INTO hello VALUES ('Fulvigula', 'quack!')"
conn = duckdb.connect(database="quack.duckdb")
conn.sql(sql)
conn.close()
```

After running the preceding cell, we have safely added this new row to our `hello` table in the `quack.duckdb` disk-based database, we created previously.

A convenient way to ensure that you close your connections is to use connection objects as **context managers**. Context managers are Python objects that can be used with the `with` keyword to automatically perform certain actions while entering and exiting the `with` statement's code block. They are commonly used for managing external resources such as database connections, network connections, and file descriptors. Here's an example of using a DuckDB connection object as a context manager:

```
sql = "INSERT INTO hello VALUES ('Labradorius', 'quack!')"
with duckdb.connect(database="quack.duckdb") as conn:
    conn.sql(sql)
```

As we can see from this example, to connect to DuckDB using a context manager, we use the `with` statement, followed by a code block that contains the database interactions we wish to perform. The `as` keyword allows us to capture the connection object as a variable, for reference inside the indented code block of the `with` statement. When the code block is exited, the connection's `close()` method will be called automatically for us. This provides a convenient way to create a new connection and also automate its closing, freeing you from having to remember to explicitly close every connection you create.

Connection objects are also closed implicitly when they go out of scope. This happens, for example, when a connection object is created inside a function body, and then the function exits (without returning the connection object). The connection object is then no longer in scope and will therefore be closed. However, rather than relying on this behavior, we recommend either using an explicit `close()` or a context manager to ensure that connections are safely closed.

Sharing disk-based databases between processes

Persistent disk-based DuckDB databases can be shared across multiple processes, provided that all databases are connected in read-only mode. You can specify that a database is opened in read-only mode by giving the `connect()` function's `read_only` parameter the value `True`:

```
conn = duckdb.connect(
    database="quack.duckdb",
    read_only=True
)
# query the database here...
conn.close()
```

Note that the `read_only` flag can only be set to `True` when connecting to a disk-based database. A read-only in-memory database would not make sense as it would never have any data that could be queried. The ability for a single database to support multiple read-only connections can be used to allow multiple consuming processes to connect to and query a single disk-based database. While it's unlikely you'll need to do this while performing data analysis in a notebook, this pattern may be useful for some specific DuckDB use cases where multiple consuming services need to be able to query a single database concurrently.

Installing and loading extensions

The Python client provides explicit methods for both installing and loading extensions, which correspond to the INSTALL and LOAD DuckDB SQL commands we saw in *Chapter 5*. These methods are invoked on connection objects, meaning that extensions are both installed and loaded on a per database-connection basis. To install and load the `spatial` extension in the default database, we run the following:

```
duckdb.install_extension("spatial")
duckdb.load_extension("spatial")
```

As we have called these methods against the `duckdb` module, the corresponding commands are dispatched to the default database. We can confirm that the `spatial` extension is installed and loaded in the default database with the following query:

```
duckdb.sql(
    """
    SELECT *
    FROM duckdb_extensions()
    WHERE loaded = true
    """
)
```

The result of this query shows us that the `spatial` extension is indeed installed and loaded. Note that you may see a slightly different set of loaded extensions depending on your environment.

extension_name varchar	loaded boolean	installed boolean	install_path varchar	description varchar	aliases varchar[]	extension_version varchar
icu	true	true	(BUILT-IN)	Adds support for tim…	[]	
jemalloc	true	true	(BUILT-IN)	Overwrites system al…	[]	
json	true	true	(BUILT-IN)	Adds support for JSO…	[]	
parquet	true	true	(BUILT-IN)	Adds support for rea…	[]	
spatial	true	true	/home/ned/.duckdb/…	Geospatial extension…	[]	8ac803e
tpcds	true	true	(BUILT-IN)	Adds TPC-DS data gen…	[]	
tpch	true	true	(BUILT-IN)	Adds TPC-H data gene…	[]	

Figure 7.6 – All loaded and installed extensions in our default database
after installing and loading the spatial extension

Let's run the same query again on a new in-memory database to confirm that the `spatial` extension was only installed and loaded for the default database:

```
conn = duckdb.connect()
conn.sql(
    """
    SELECT *
    FROM duckdb_extensions()
    WHERE loaded = true
    """
)
```

This query shows that the new in-memory database does not have the `spatial` extension installed or loaded:

extension_name varchar	loaded boolean	installed boolean	install_path varchar	description varchar	aliases varchar[]	extension_version varchar
icu	true	true	(BUILT-IN)	Adds support for time zones …	[]	
jemalloc	true	true	(BUILT-IN)	Overwrites system allocator …	[]	
json	true	true	(BUILT-IN)	Adds support for JSON operat…	[]	
parquet	true	true	(BUILT-IN)	Adds support for reading and…	[]	
tpcds	true	true	(BUILT-IN)	Adds TPC-DS data generation …	[]	
tpch	true	true	(BUILT-IN)	Adds TPC-H data generation a…	[]	

Figure 7.7 – All loaded and installed extensions in a newly created connection in-memory database, without installing or loading any extensions

To install and load this extension in our new database, we would need to run the following:

```
conn.install_extension("spatial")
conn.load_extension("spatial")
```

This also serves as a reminder that multiple DuckDB databases running in a Python process are distinct and do not share state with each other.

Summary

In this chapter, we moved beyond using the simple DuckDB CLI for working with DuckDB databases and began our journey into using DuckDB with Python. We started by getting our environment set up with the necessary dependencies. We then looked at the two different ways to connect to DuckDB from Python: using the default in-memory database, which is convenient for performing interactive data analysis and other types of ad-hoc DuckDB usage, and the explicit use of connection objects, which offers increased flexibility within our DuckDB-powered workflows, as well as being the safer way to build Python libraries that interact with DuckDB databases. We then saw how we can connect to disk-based DuckDB databases, allowing us to persist prepared databases for subsequent use, before coving the importance of closing database connections, especially when using disk-based databases. We also looked at the more specialized use case of multiple processes connecting to a disk-based database in read-only mode, before finishing up with how to install and load DuckDB extensions in specific databases when using the Python client.

Now that our environment is set up for working with DuckDB in Python, and we are familiar with the different ways of connecting to DuckDB databases, we're ready to dive deeper into using DuckDB with Python. In the next chapter, we'll work our way through the different features of the DuckDB Python client, which will prepare us for making effective use of DuckDB for analytical workflows and developing software that integrates with DuckDB.

8

Exploring DuckDB's Python API

In the previous chapter, we started our journey into using DuckDB and Python by getting the DuckDB Python client installed into a Jupyter Notebook-based environment, as well as unpacking the different ways to connect to DuckDB databases. In this chapter, we'll continue where we left off by diving into a practically oriented walkthrough of the core components of the DuckDB Python API, which will enable you to hit the ground running working with DuckDB in Python.

More specifically, we'll explore the following aspects of working with DuckDB in Python:

- Interacting with DuckDB using the Relational API
- Interacting with DuckDB using the Python DB-API
- Consuming from and converting to Python in-memory data formats, including Apache Arrow tables, pandas Dataframes, and Polars Dataframes
- How Python types are converted to DuckDB types
- Using Python functions in DuckDB queries
- Exception handling with DuckDB

By the end of this chapter, you'll have all the ingredients you need to bring DuckDB into your Python data analysis workflows as well as to develop DuckDB-powered tools and applications in Python.

Technical requirements

The exercises in this chapter require that you have a Python virtual environment set up with the necessary dependencies installed and also have a Jupyter Notebook-compatible IDE set up. If you have not done so already, please follow the instructions in the *Technical requirements* section of *Chapter 7* before jumping into the examples in this chapter.

GitHub repository

You will find data and the code examples covered in this chapter in the chapter_08 folder in the book's GitHub repository:

https://github.com/PacktPublishing/Getting-Started-with-DuckDB/tree/main/chapter_08.

Obtaining the dataset

To help us unpack DuckDB's Relational API for Python, we're going to explore the *Seattle Pet Licenses* dataset, which contains information about pet licenses that have been registered in the city of Seattle. We'll be performing a small amount of **exploratory data analysis (EDA)** over this dataset, which will give us an opportunity to put DuckDB's Python client through its paces. We have included a snapshot of this dataset in this chapter's directory in the GitHub repository. This is the dataset that was used to generate the output you see in this chapter.

Alternatively, you can download the most recent version of the dataset from the Seattle Open Data portal: https://data.seattle.gov/Community/Seattle-Pet-Licenses/jguv-t9rb. You'll need to select the **CSV** option from the **Export** button. This will download the CSV file you'll need. The file that was used to produce the output shown in this chapter was Seattle_Pet_Licenses_20240505.csv, corresponding to the current edition of this dataset at the time of writing. You may fetch a more recent version of the dataset when you follow these steps, which could include subsequent pet licenses that have been issued. This means that your results might look slightly different from those we have provided. We've also renamed this CSV file Seattle_Pet_Licenses.csv. You'll either need to rename the file you downloaded accordingly or adjust the CSV queries in the examples to target the file you downloaded.

Working with the Relational API

DuckDB's Relational API provides a convenient and effective Python interface for composing and working with DuckDB queries. It provides a flexible and efficient way to interact with DuckDB in Python and is especially well suited to interactive data analysis workflows.

The Relational API revolves around DuckDBPyRelation objects, which are more generally referred to as *relations*. You can think of a DuckDB relation as a representation of a DuckDB query. Relations do not contain data themselves but rather contain all the information about a query required for execution. Relations are lazily evaluated, which means that when you are working with them, nothing is run against the database until the result set for a query is needed, such as when you want to display their results interactively or when the full result set is materialized for loading into a table or exporting to a specific data format.

The laziness of DuckDB's relations provides a number of benefits. Firstly, when constructing a query as a relation in an interactive session, displaying a relation will result in DuckDB streaming in only the first 10,000 rows. This helps support fast query-development feedback loops, as you don't need to wait for your full query to run before being able to see results. Additionally, the Relational API allows you to iteratively construct queries by applying consecutive operations against relation objects. Only when the results of the relation object are needed will the query be executed against the database. When composing queries as relations this way, DuckDB's query planner is able to apply more query optimizations than would be possible if you were sequentially executing individual queries, which would see the results being fully materialized after each query.

In this section, we'll cover the following:

- How to create relations from SQL queries, files, and tables in your database

- How to query relations using SQL queries

- How to perform data transformations using relations

- How to export the results of relations to disk

- How to load and insert the contents of relations into tables

Creating relations

We'll start our exploration of the Relational API by looking at some of the ways you can construct relations. The most general-purpose way to create a relation object is with the `DuckDBPyConnection.sql()` method. This takes a SQL query as a string and returns a corresponding relation object representing your query. A relation object is always associated with a specific DuckDB database, which is why the `sql()` method is always called on a connection object. As we saw in *Chapter 7*, calling the convenient `duckdb.sql()` method will construct a relation using the connection to the default in-memory database, or you can construct relations using explicitly defined connection objects; for example, `conn.sql()`.

Let's start by creating a simple relation that uses the default database. We'll use DuckDB's `pi` function, which returns the value of the mathematical constant π. As with all the examples in this chapter, you should now run this Python code in a notebook cell:

```
import duckdb
pi_relation = duckdb.sql("SELECT pi() AS pi")
type(pi_relation)
```

The output of this cell shows us the Python type of the value assigned to the `pi_relation` variable:

```
duckdb.duckdb.DuckDBPyRelation
```

Here, we've created a relation object that represents our simple query and have assigned it to the `pi_relation` variable. We've also used Python's `type()` function to show that the object returned by the `sql()` method is indeed a DuckDB relation. Recall from *Chapter 7* that the final expression in the cell of a Jupyter notebook will be displayed on execution, so we see the following after executing this notebook cell:

```
duckdb.duckdb.DuckDBPyRelation
```

At this point, we have not yet executed anything against the default database; we've simply created our relation object. One of the ways a relation will be executed is by displaying its results, which we can do with the `DuckDBPyRelation.show()` method:

```
pi_relation.show()
```

Our relation has now been run against the default database, giving us the following output:

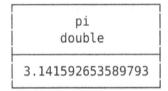

Figure 8.1 – The result of displaying a simple relation object

A convenient feature of relation objects is that their default output representation is produced by automatically calling the `show()` method. This means you can omit the `show()` method call when the expression that produces a relation object occurs at the end of a code cell in your notebook, as your IDE will automatically display its representation as output. We'll be taking advantage of this convenience in the examples in this chapter. You can try this now by simply evaluating the `pi_relation` variable in a notebook cell:

```
pi_relation
```

Note that the `sql()` method will only return a relation object when the SQL query contains a SELECT statement. Other types of SQL statements are executed immediately against the database without returning a relation since no result set needs to be returned.

Creating relations from SQL queries

Let's move on now to something a little more interesting, by using the `sql()` method to load our dataset from the `Seattle_Pet_Licenses.csv` CSV file. To load this file, we'll construct a SQL query that uses DuckDB's `read_csv` function, as we covered in *Chapter 2*:

```
duckdb.sql(
    """
```

```
    SELECT *
    FROM read_csv('Seattle_Pet_Licenses.csv')
    """
)
```

As our call to `duckdb.sql()` is the only expression in this cell, it is evaluated, and the resulting relation object is displayed:

License Issue Date varchar	License Number varchar	Animal's Name varchar	Species varchar	Primary Breed varchar	Secondary Breed varchar	ZIP Code varchar
December 18 2015	S107948	Zen	Cat	Domestic Longhair	Mix	98117
June 14 2016	S116503	Misty	Cat	Siberian	NULL	98117
August 04 2016	S119301	Lyra	Cat	Mix	NULL	98121
February 13 2019	962273	Veronica	Cat	Domestic Longhair	NULL	98107
August 10 2019	S133113	Spider	Cat	LaPerm	NULL	98115
November 21 2019	8002549	Maxx	Cat	American Shorthair	NULL	98125
May 24 2020	S142869	Mickey	Cat	Domestic Longhair	NULL	98126
July 03 2020	S112835	Diamond	Cat	Domestic Shorthair	Mix	98103
July 21 2020	S131986	Nacho	Cat	Domestic Shorthair	Mix	98126
August 18 2020	8019541	Gracie	Cat	Domestic Medium Hair	Mix	98133
·	·	·	·	·	·	·
·	·	·	·	·	·	·
·	·	·	·	·	·	·
November 22 2023	S148106	Gibbs	Cat	Domestic Shorthair	Mix	98116
November 22 2023	8042587	Barry	Cat	American Shorthair	Mix	98109
November 22 2023	8042588	Penny	Cat	American Shorthair	Mix	98109
November 22 2023	8050533	Sabine	Cat	Domestic Medium Hair	York Chocolate	98144
November 22 2023	8050534	Milo	Cat	Domestic Medium Hair	Mix	98144
November 22 2023	S137987	Spike	Cat	Domestic Medium Hair	NULL	98117
November 22 2023	S137986	Scout	Cat	Domestic Shorthair	NULL	98117
November 22 2023	8013084	Honeybee	Cat	American Shorthair	Mix	98106
November 22 2023	S157559	Bug	Cat	Ragdoll	Siamese	98144
November 22 2023	8042668	Beerus	Cat	Domestic Shorthair	Mix	98125

? rows (>9999 rows, 20 shown)   7 columns

Figure 8.2 – The result of displaying the relation created via a SQL query that loads the dataset's CSV file

We can see from this output that DuckDB has executed our query, and the CSV reader has loaded records from our CSV that have seven columns. We can also see that the full contents of the CSV file have not been consumed, with only the first 10,000 rows having been fetched. This is the laziness of DuckDB's relations in action, where batches of rows are streamed through as needed. This allows us to rapidly preview the initial results from executing relations, even when their underlying result set contains massive numbers of rows—a convenient feature when prototyping loading and transformation logic. DuckDB is able to work out when it needs to fully materialize all results of a query, such as when converting query results into pandas Dataframes or exporting results to a file on disk. We'll see both of these scenarios soon.

Creating relations from files

The Relational API's `sql()` method is a general-purpose way to create relations and interact with DuckDB databases, giving you full access to DuckDB's SQL interface. While you can use DuckDB's file reading functions via SQL queries supplied to the `sql()` method—as just saw—the Relational API also offers a range of methods that provide deeper integration with the Python language and that can often be more convenient to work with. Notably, the Relational API provides convenience methods for loading data from CSV, JSON, and Parquet files via the following methods:

- `DuckDBPyConnection.read_csv()`
- `DuckDBPyConnection.read_parquet()`
- `DuckDBPyConnection.read_json()`

Just as with the `sql()` method, these are all exposed via connection objects, meaning that they can be called against the `duckdb` module—which will use the default database—and they can be called against explicitly created connection objects. They all take a target file path for reading as a required argument and return a relation object, which is a lazily evaluated representation of the extracted contents of their target files.

We'll use the `read_csv()` method via the default database to create a new relation object that represents the contents of the data loaded from the CSV file:

```
pets_csv_relation = duckdb.read_csv(
    "Seattle_Pet_Licenses.csv"
)
```

The resulting relation object assigned to the `pets_csv_relation` variable is logically equivalent to the relation we produced earlier using the `sql()` method, but rather than writing our query in SQL, we've used a Relational API method to construct it. Note that, since we haven't displayed the results of this relation yet, nothing has been executed yet.

You may have noticed earlier, when we used the `sql()` method to load our CSV file, that the automatically derived schema of the records that came back from our CSV file is not quite right; the first column in the `License Issue Date` schema should have a data type of `DATE` rather than `VARCHAR`. We can see this is still an issue for the relation we just created with the `read_csv()` method by inspecting its `types` property:

```
pets_csv_relation.types
```

This gives us a Python list of the column-data types for this relation, which shows us the first column is a VARCHAR data type:

```
[VARCHAR, VARCHAR, VARCHAR, VARCHAR, VARCHAR, VARCHAR, VARCHAR]
```

As we saw in *Chapter 2*, DuckDB's CSV reader has parameters that allow us to specify configuration for how to read target files. The Relational API's `read_csv()` method accepts corresponding parameters, which we can use to load the `License Issue Date` column as a DATE type. We'll need to use the `dtype` keyword argument to indicate this column is a DATE type, as well as the `date_format` keyword argument to specify how these text values should be parsed into dates. The format specifiers that can be used in the `date_format` string are documented in the DuckDB documentation: `https://duckdb.org/docs/sql/functions/dateformat.html`. Here's what our new `read_csv()` call looks like:

```
pets_csv_relation = duckdb.read_csv(
    "Seattle_Pet_Licenses.csv",
    dtype={"License Issue Date": "DATE"},
    date_format="%B %d %Y",
)
pets_csv_relation.types
```

We can see from the `types` attribute of this new relation that the first column now indeed has a DATE type:

```
[DATE, VARCHAR, VARCHAR, VARCHAR, VARCHAR, VARCHAR, VARCHAR]
```

Let's quickly peek at our data. Just as you might throw a LIMIT clause onto a SQL query to only look a small subset of query results, we can call the `DuckDBPyRelation.limit()` method to do the same thing:

```
pets_csv_relation.limit(5)
```

Evaluating this shows us the first five rows yielded by this relation:

| License Issue Date | License Number | Animal's Name | Species | Primary Breed | Secondary Breed | ZIP Code |
date	varchar	varchar	varchar	varchar	varchar	varchar
2015-12-18	S107948	Zen	Cat	Domestic Longhair	Mix	98117
2016-06-14	S116503	Misty	Cat	Siberian	NULL	98117
2016-08-04	S119301	Lyra	Cat	Mix	NULL	98121
2019-02-13	962273	Veronica	Cat	Domestic Longhair	NULL	98107
2019-08-10	S133113	Spider	Cat	LaPerm	NULL	98115

Figure 8.3 – A relation containing the first five rows from our parsed CSV file, which has the correct type for the License Issue Date column

To see the full set of parameters that the Relational API's `read_csv()` method exposes, you can call Python's built-in `help` function on it:

```
help(duckdb.read_csv)
```

You can do this also for the analogous `read_parquet()` and `read_json()` methods, as well as consulting the full Python API reference documentation for all methods in the DuckDB Python API: `https://duckdb.org/docs/api/python/reference`.

It's worth bearing in mind that the Relational API is essentially a Python interface to the same underlying DuckDB internal interface that DuckDB SQL targets. You can, in general, use SQL queries to achieve the same results as Relational API method calls, and you can mix and match these in your workflows, depending on your needs and preferences. Relation objects have a convenient `sql_query()` method that returns a string containing the SQL query that is equivalent to the relation:

```
print(pets_csv_relation.sql_query())
```

This gives us the following DuckDB SQL query:

```
SELECT * FROM read_csv_auto(['Seattle_Pet_Licenses.csv'], (auto_
detect = false), (dtypes = {'License Issue Date': 'DATE'}), (delim
= ','), ("quote" = '"'), (null_padding = false), ("escape" = '"'),
(dateformat = '%B %d %Y'), (all_varchar = false), ("columns" =
{'License Issue Date': 'DATE', 'License Number': 'VARCHAR', 'Animal's
Name': 'VARCHAR', 'Species': 'VARCHAR', 'Primary Breed': 'VARCHAR',
'Secondary Breed': 'VARCHAR', 'ZIP Code': 'VARCHAR'}), (max_line_size
= 2097152), (normalize_names = false), ("header" = true), ("skip" =
0), ("parallel" = true), (maximum_line_size = 2097152))
```

You may have noticed that the `read_csv` SQL function call has a range of parameters specified that we did not provide any configuration for in our Python `read_csv()` method call. DuckDB has used the CSV sniffer—which we discussed in *Chapter 4*—to infer appropriate CSV reader parameters for this CSV file, and then it has generated a query with these filled in explicitly, while incorporating the configuration we did specify, and also disabling auto-detection.

Creating relations from tables

We can also create relations from existing tables in a DuckDB database. Since we haven't created any tables in our in-memory database yet, we'll start by making one with the `DuckDBPyRelation.to_table()` method. When called on its relation object, this method will create a new table with the provided name and populate it with the results of executing the relation:

```
pets_csv_relation.to_table("seattle_pets_dataset")
```

Running a SHOW TABLES statement on our database shows us that our table has been created and added to the database catalog:

```
duckdb.sql("SHOW TABLES")
```

We get the following result:

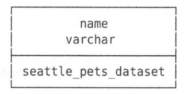

name
varchar
seattle_pets_dataset

Figure 8.4 – Output of the SHOW TABLES statement

Now, we can make a relation from this table we've just created using the DuckDBPyConnection.table() method:

```
pets_table_relation = duckdb.table("seattle_pets_dataset")
```

We can now use this relation to work with this table via the Relational API. For example, the DuckDBPyRelation.describe() method is equivalent to issuing a SQL DESCRIBE statement:

```
pets_table_relation.describe()
```

We get the following output:

aggr varchar	License Issue Date varchar	License Number varchar	...	Primary Breed varchar	Secondary Breed varchar	ZIP Code varchar
count	42567	42567	...	42567	28373	42440
mean	NULL	NULL	...	NULL	NULL	NULL
stddev	NULL	NULL	...	NULL	NULL	NULL
min	2015-10-21	015352	...	Abruzzese Mastiff	Abruzzese Mastiff	22308
max	2024-04-05	s124879	...	Xoloitzcuintli	York Chocolate	V5J 1P8
median	NULL	NULL	...	NULL	NULL	NULL
6 rows					8 columns (6 shown)	

Figure 8.5 – Output of applying the describe() method to a relation

Another convenient method the Relational API makes available is DuckDBPyRelation.count(), which is equivalent to applying a count(*) aggregation operation to the query in the relation object:

```
pets_table_relation.count("*")
```

Evaluating the relation that this expression produces gives us the number of rows in the `seattle_pets_dataset` table:

```
┌──────────────┐
│ count_star() │
│    int64     │
├──────────────┤
│        42567 │
└──────────────┘
```

Figure 8.6 – The number of rows in the seattle_pets_dataset table

Before we move on, we'll remove the table we just created. Not every DuckDB SQL statement is exposed via methods in the Relational API, which is why we used the `sql()` method earlier to run the SHOW TABLES SQL statement. At the time of writing, the Relational API does not have a method corresponding to the SQL DROP statement, so we'll use SQL:

```
duckdb.sql("DROP TABLE seattle_pets_dataset")
```

Having looked at different ways to create relations, next, we'll see how the Python DuckDB client enables convenient querying of relation objects.

Querying relations

When running in a Python process, DuckDB can query existing relation objects as if they were database tables. When writing a SQL query, this means you can invoke the names of a variable containing a relation object that is currently in scope in your Python session, and DuckDB will be able to query it as though it were a table in your database's catalog.

Let's use this feature to query the relation assigned to the `pets_table_relation` variable we created previously. We'll construct a query that selects a subset of the columns derived from our CSV file and also rename them, which will make our dataset easier to work with for performing data analysis. To do this, we'll write an appropriate SQL query, specifying the `pets_csv_relation` variable in the FROM clause:

```
pets_cleaned_relation = duckdb.sql(
    """
    SELECT
        "License Issue Date" AS issue_date,
        "Animal's Name" AS pet_name,
        "Species" AS species,
        "Primary Breed" AS breed
    FROM pets_csv_relation
    """
```

```
)
pets_cleaned_relation.limit(5)
```

Note that we've used a triple-quoted Python string for our query, which allows us to include new lines and also means we don't have to worry about escaping any quote characters in our SQL query. When we run this cell, it produces a relation that, when displayed, shows us five rows loaded from the CSV file with our cleaning steps applied:

issue_date date	pet_name varchar	species varchar	breed varchar
2015-12-18	Zen	Cat	Domestic Longhair
2016-06-14	Misty	Cat	Siberian
2016-08-04	Lyra	Cat	Mix
2019-02-13	Veronica	Cat	Domestic Longhair
2019-08-10	Spider	Cat	LaPerm

Figure 8.7 – A relation containing the first five rows from our cleaned pet registrations dataset

This convenient feature of DuckDB is known as **replacement scanning**. It enables DuckDB to lookup alternative data sources when a target table name does not occur in the database catalog. In this case, the pets_cleaned_relation table identifier is not present in our database's catalog. This causes DuckDB to attempt a replacement scan, identifying a variable with the same name in our Python session's global scope, which contains a DuckDBPyRelation object, meaning that the replacement scan will succeed and this relation with be used as the data source for this query. As we'll see later in this chapter, in addition to relation objects, replacement scans can also be used to query pandas Dataframes, Polars Dataframes, and Apache Arrow tables.

Let's do this again, this time querying the new relation we just created, using it to retrieve the earliest and latest pet license issue dates:

```
min_max_relation = duckdb.sql(
    """
    SELECT min(issue_date), max(issue_date)
    FROM pets_cleaned_relation
    """
)
min_max_relation
```

Evaluating this gives us the following output:

min(issue_date) date	max(issue_date) date
2015-10-21	2024-04-05

Figure 8.8 – A relation containing the earliest and latest license-issued dates in the dataset

It's worth highlighting that this SQL query has involved two replacement scans. It invokes pets_ cleaned_relations as a data source, which itself contains a query that leverages a replacement scan, invoking pets_csv_relation as a data source.

This illustrates how replacement scans allow us to incrementally build queries through intermediate relation objects while also separating out logical groups of operations into distinct Python expressions. Importantly, since DuckDB relations are lazily evaluated, we can use this technique to build up efficient data processing pipelines, enabling deferred execution until our ultimate relation object needs to be displayed or fully materialized, such as when loading into a table or exporting to a file. Chaining relation-object queries like this is analogous to SQL queries with chained references to database views; however, rather than being defined in our database's catalog, we are managing each node in the pipeline within our Python session through relation objects. It's also worth pointing out that this contrasts with using a data processing tool such as pandas to define a data pipeline within a Python session, where each intermediate result would be fully materialized, regardless of whether it is subsequentially used.

To further underscore the lazy nature of the pipeline we have created from relation objects, let's call the DuckDBPyRelation.explain() method on the relation object that contains our aggregate query:

```
print(min_max_relation.explain())
```

This is equivalent to using the SQL EXPLAIN statement on the query that this relation represents. Evaluating it gives us the following query plan for this relation:

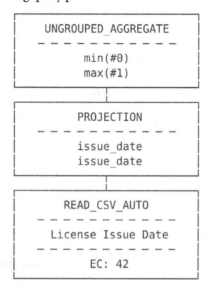

Figure 8.9 – The output from applying the explain() method to a relation, showing its query plan

Even though we have used three distinct Relational API calls to generate this relation object (our original read_csv() call, followed by two sql() queries leveraging replacement scans), deferred execution has enabled DuckDB to consolidate this into a single optimized database query that contains three operations:

- Reading the initial CSV file
- Retrieving only the renamed issue_date column from this CSV file (a *projection* operation)
- Performing two aggregation operations over this single column

Notably, because our final query only required a single column from the CSV file, License Issue Date, the query optimizer is able to use a projection pushdown, only loading the contents of this column when reading the CSV.

In the next section, we'll continue to explore how the Relational API allows us to construct efficient DuckDB queries in Python.

Transformations with relations

DuckDB's Relational API is highly flexible and can be used in several different modes to build queries that define data transformations. Here, we will go through some different modes of use to help provide you with a framework for thinking about how to use the Relational API for your own workflows. We'll use the relation object we created previously, `pets_cleaned_relation`, as the data source for our transformations, building up new relation objects from it, representing queries containing both our original data cleaning steps and any subsequent transformations.

Comparing SQL queries and Relational API method calling

We've seen how DuckDB's Relational API allows us to create DuckDB relations using SQL queries using its `DuckDBPyConnection.sql()` method and via other methods exposed by connection objects that don't involve SQL, such as `DuckDBPyConnection.read_csv()` and `DuckDBPyConnection.table()`. This alternative between using SQL queries and Relational API method calls forms something of a theme when working with the Relational API in the Python client, which we'll continue to see as we look at how to perform transformations using relations.

Broadly speaking, when you want to apply a transformation to a relation object, you can do the following:

- Define your transformations using SQL via the `sql()` method, optionally leveraging the convenience of replacement scans to query other relation objects, which allows you to compose your transformations into pipelines of multiple SQL queries

- Leverage `DuckDBPyRelation` methods that apply transformation operations and return new relation objects, enabling the composition of queries via method chaining, without the use of SQL

You can adopt either approach when querying DuckDB in Python, depending on your needs and preferences. These different styles of query creation can also be mixed and matched; you don't need to adopt one exclusively.

We'll now compare these two approaches with a simple query involving a transformation. Let's say we want to find all records for pet licenses given to pigs. Starting with the SQL-based approach, to perform this filter operation, we'd naturally reach for a `WHERE` clause. We can create a relation that represents this query by passing an appropriate SQL query to the `sql()` method, which queries the `pets_cleaned_relation` variable using a replacement scan:

```
duckdb.sql(
    """
    SELECT *
    FROM pets_cleaned_relation
    WHERE species = 'Pig'
    """
)
```

This shows us two pigs in this dataset, Millie and Calvin, who have registered pet licenses:

issue_date date	pet_name varchar	species varchar	breed varchar
2022-11-03	Millie	Pig	Pot-Bellied
2023-06-01	Calvin	Pig	Pot-Bellied

Figure 8.10 – A relation containing two records for registrations of pigs in the dataset

We can also generate a relation representing this same query through the `DuckDBPyRelation.filter()` method. Calling this method on a relation object will return a new relation with a filter operation applied to the relation, using the argument passed to the method as the filter expression. Applying this approach to our data-source relation object, we have the following:

```
pets_cleaned_relation.filter("species = 'Pig'")
```

While we didn't use a complete SQL query to create this relation, we did use a SQL expression as an argument to the `filter()` method—the same expression we used for the `WHERE` clause for our SQL query. Using SQL expressions with Relational API method calls enables capturing data transformations as concise Python expressions, which is particularly useful for performing rapid data analysis. Representing SQL expressions as strings does have the drawback of offering poorer integration with Python: they do not lend themselves to being composed programmatically, and they cannot support validation for subcomponents of expression, limiting opportunities for localized diagnostic feedback on errors.

The Expression API

DuckDB's Python client provides an Expression API that supports the dynamic construction of Python expression objects, enabling the flexible composition of complex relation objects without the need for any SQL strings. Let's see how we can use this to construct an equivalent relation object to our last query, this time using an expression object as an argument to the `filter()` method.

Our filter expression involves checking values from the `species` column for equality with the `'Pig'` string literal. To reference these components, we'll need to construct instances of the Expression API's `ColumnExpression` and `ConstantExpression` classes, respectively, which we can then use to create the expression we need with Python's equality operator:

```
from duckdb import ColumnExpression, ConstantExpression

species_col = ColumnExpression("species")
pig_constant = ConstantExpression("Pig")
pets_cleaned_relation.filter(species_col == pig_constant)
```

On evaluating and displaying the resulting relation object, you'll see that we've constructed the same query once again that finds our two pigs.

We can also compose expression objects to define complex constraints. If, for example, we wanted to find all cats with the name Leeloo, we could do this using a SQL expression like so:

```
pets_cleaned_relation.filter(
    "species = 'Cat' AND pet_name = 'Leeloo'"
)
```

It produces the following result when displayed:

issue_date date	pet_name varchar	species varchar	breed varchar
2022-04-13	Leeloo	Cat	Domestic Shorthair
2022-08-15	Leeloo	Cat	Domestic Shorthair
2022-08-21	Leeloo	Cat	Domestic Shorthair
2022-10-20	Leeloo	Cat	American Shorthair

Figure 8.11 – The results from a query finding all cats named Leeloo

Here's how we can create the same relation using an expression object as the argument to the `filter()` method:

```
name_col = ColumnExpression("pet_name")
is_cat = species_col == ConstantExpression("Cat")
is_leeloo = name_col == ConstantExpression("Leeloo")
pets_cleaned_relation.filter(is_cat & is_leeloo)
```

Note that we created two reusable expression objects, each capturing one of our constraints. We then combined then converted them using Python's & operator to create an expression object representing our composite constraint. This illustrates how the Expression API provides deeper Python integration than SQL expressions, giving us the flexibility of programmatically building up complex expressions.

Method chaining

We have seen how replacement scanning enables the incremental building of queries as relation objects. Another way the Relational API enables incremental query construction is via method chaining. Relational API methods that apply an operation to an existing relation all return a new relation that represents the updated query. The resulting relation object can have subsequent operations applied by further method calls. This allows us to build complex queries using chains of relational method calls.

Let's see an example:

```
pets_cleaned_relation.filter(is_cat).limit(5)
```

It gives the following result:

issue_date date	pet_name varchar	species varchar	breed varchar
2015-12-18	Zen	Cat	Domestic Longhair
2016-06-14	Misty	Cat	Siberian
2016-08-04	Lyra	Cat	Mix
2019-02-13	Veronica	Cat	Domestic Longhair
2019-08-10	Spider	Cat	LaPerm

Figure 8.12 – The results from finding all cat licenses and limiting to five rows

In this example, we started with our `pets_cleaned_relation` variable, which contains our CSV loading and cleaning operations. We then applied a `filter()` method, giving it the `is_cat` expression we created previously as an argument, which will filter the query results down to only pets with a `species` value of `Cat`. We then applied the `DuckDBPyRelation.limit()` method with a value of 5, which has the same effect as applying a `LIMIT 5` clause to the query our relation object now holds.

Here's a further example of method chaining, whereby we create a relation containing a query that finds the number of distinct cat names in the dataset. Note that we've wrapped this Python expression with parentheses so that we can format each element in the chain on a new line:

```
num_cat_names_rel = (
    pets_cleaned_relation
    .filter(is_cat)
    .select("pet_name")
    .distinct()
    .count("*")
)
num_cat_names_rel
```

We get the output as follows:

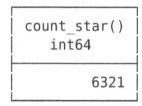

Figure 8.13 – The number of distinct cat names in the dataset

In this chain, we followed the filtering step with a `DuckDBPyRelation.select()` call, passing it the `"pet_name"` string so that our query now only extracts this column. We then applied the `DuckDBPyRelation.distinct()` method, which modifies our query to now only select unique `pet_name` values, and then finally called `DuckDBPyRelation.count()`, resulting in our final relation object giving us the number of unique cat names in the dataset.

This illustrates the power of iteratively composing queries through a series of method calls that are applied to relation objects. This is made possible by DuckDB's Relational API being designed to support method chaining, where each method returns a new relation that incorporates the results of the operation just applied. This is reminiscent of the pandas API, in which subsequent operations can be performed on Dataframes by chaining method calls. A notable difference, however, is that internally, pandas will fully evaluate and materialize the results of each operation before invoking the next method on the results of the previous. DuckDB's Relational API is designed around lazy evaluation, deferring execution as late as possible. This means that all operations can be combined and converted into a final query before it is executed. This provides DuckDB with more opportunities for query optimization and the risk of hitting memory limitations during the calculation of intermediate operations.

The flexibility of the Relational API means that there are often different ways to compose equivalent queries. For example, we can create the same query from before, but rather than using the `select()` and `distinct()` methods, we can replace them with `DuckDBPyRelation.unique()`, which takes a column expression for deduplication to be performed on:

```
alt_num_cat_names_rel = (
    pets_cleaned_relation
    .filter(is_cat)
    .unique("pet_name")
    .count("*")
)
alt_num_cat_names_rel
```

You might like to try calling `explain()` on both `num_cat_names_rel` and `alt_num_cat_names_rel`, which will show you that the query plans DuckDB generates for these relations are the same.

It can also be helpful to inspect the SQL query that corresponds to a complex relation object. Let's use the `sql_query()` method on the `alt_num_cat_names_rel` relation:

```
num_unique_cats_sql = alt_num_cat_names_rel.sql_query()
print(num_unique_cats_sql)
```

This prints out the following SQL query:

```
SELECT count_star() FROM (SELECT DISTINCT pet_name FROM (SELECT *
FROM (SELECT "License Issue Date" AS issue_date, "Animal's Name"
AS pet_name, Species AS species, "Primary Breed" AS breed FROM
pets_csv_relation) AS unnamed_relation_0a156a5f3dc52df2 WHERE
(species = 'Cat')) AS unnamed_relation_0a156a5f3dc52df2) AS unnamed_
relation_0a156a5f3dc52df2 GROUP BY ALL
```

Since this SQL is a little hard to read, we'll use the `sqlparse` library (which we installed via our `requirements.txt` file in *Chapter 7*) to format it into something easier to read:

```
import sqlparse
formatted_sql = sqlparse.format(
    num_unique_cats_sql,
    reindent=True
)
print(formatted_sql)
```

This prints out the following much more cleanly formatted SQL:

```
SELECT count_star()
FROM
   (SELECT DISTINCT pet_name
    FROM
       (SELECT *
        FROM
           (SELECT "License Issue Date" AS issue_date,
                   "Animal's Name" AS pet_name,
                   Species AS species,
                   "Primary Breed" AS breed
            FROM pets_csv_relation) AS unnamed_relation_0a156a5f3dc52df2
        WHERE (species = 'Cat')) AS
           unnamed_relation_0a156a5f3dc52df2) AS
             unnamed_relation_0a156a5f3dc52df2
GROUP BY ALL
```

If you look through this query closely, you'll see that not only does it contain the SELECT statement that defines our `pets_cleaned_relation` relation that uses the `pets_csv_relation` variable as a data source, but it also includes all of the operations that were applied via our relation method-call chaining.

Putting it all together

Before we finish looking at using the Relational API to apply transformations, we'll look at one last query, comparing the approach of defining our query with SQL and constructing it using Relational API method calls and expression objects.

Let's say we wanted to find the 10 pets in the dataset with the longest names, also returning the number of characters in their names. We'll start by using the `sql()` method to create a relation that represents this query with SQL:

```
duckdb.sql(
    """
    SELECT *, length(pet_name) AS name_length
    FROM pets_cleaned_relation
    ORDER BY name_length DESC
    LIMIT 10
    """
)
```

In this query, we created a new `name_length` column, which is derived by applying DuckDB's `length` function to get the number of characters for each `pet_name` value, selecting it along with all existing rows. We also used an `ORDER BY` clause to sort our results by the new `name_length` column in descending order, limiting our results to the top 10 with a `LIMIT` clause. Evaluating this cell in our notebook shows us the results:

issue_date date	pet_name varchar	species varchar	breed varchar	name_length int64
2023-08-04	Nuit Ahathoor Hecate Sappho Jezebel Lilith Crowley	Dog	Chihuahua, Short Coat	50
2024-02-10	Lady Kassandra Yu Countess of Wallingford DBE	Dog	Border Collie	45
2024-02-01	KingKing SirBeastmodeEsquire Stella Jr Sr II	Dog	Terrier, American Pit Bull	44
2024-01-09	Alyeska Juniper Cocoa Luna Taber O'Kelley	Cat	Domestic Shorthair	41
2024-01-22	WINTERDAWG PRINCESS BUTTERCUP FROSTY JANE	Dog	Alaskan Malamute	41
2022-12-01	Cascade Mountain's Out of a Dream (BRIA)	Dog	Retriever, Golden	40
2023-11-01	Houlene "Hula" Elizabeth Honeybee Jacobs	Dog	Coonhound, Bluetick	40
2023-02-10	Agnes "Aggie" Snowball Pineapple Clark	Dog	Retriever, Labrador	38
2024-02-01	Snapdragon Caylee Reid Gramila O'Keefe	Dog	Terrier, Jack Russell	38
2023-02-07	Moose Donut Franswa Cuddlewuddle Post	Dog	Bulldog, French	37
10 rows				5 columns

Figure 8.14 – The 10 longest pet names in the dataset with an additional
derived column containing the character length of each name

Now, let's see how we can compose this same query without SQL, purely using Relational API method calls and expression objects. We'll break this down into two steps. Firstly, we'll create all the expression objects we'll need:

```
from duckdb import (
    ColumnExpression,
    FunctionExpression,
    StarExpression
)

star = StarExpression()
name_col = ColumnExpression("pet_name")
name_length_col = (
    FunctionExpression("length", name_col)
    .alias("name_length")
)
name_length_sort = ColumnExpression("name_length").desc()
```

Let's go through these four expression objects we've created:

- The `StarExpression` instance assigned to `star` will allow us to select all columns in the source relation, corresponding to * in the `SELECT` clause of our SQL query.

- The `ColumnExpression` instance assigned to `name_col` will be used to reference the `pet_name` column when deriving the new `name_length` column.

- The expression object that defines our derived column is assigned to `name_length_col`, which we created using the `FunctionExpression` class, supplying it with the name of the DuckDB function we want to apply, as well as the column we want to apply it to, which is captured by the `name_col` expression object. We also use the `Expression.alias()` method of the resulting function expression to give it the `name_length` alias.

- The `name_length_sort` expression object is a reference to the `name_length` column and is defined as being in descending order via the `Expression.desc()` method. This will allow us to specify the sort order we need for our query.

With these expression objects created, we can construct our desired relation object using a concise method call chain:

```
longest_names_relation = (
    pets_cleaned_relation
    .select(star, name_length_col)
    .sort(name_length_sort)
    .limit(10)
)
```

In this method chain, we did the following:

- Used the `select()` method, whose expression arguments correspond to columns we want to retrieve. The `star` argument configures our query to return all in the base relation, and the `name_length_col` argument adds our derived column to the results.

- Applied the `DuckDBPyRelation.sort()` method, giving it the expression object we defined to specify the sort order for our query. In the SQL query we created earlier, this call corresponds to applying the `ORDER BY name_length DESC` clause.

- Finished with the `limit()` method to restrict our results to the first 10 in the ordering.

As an exercise, you may find it interesting to call both the `sql_query()` and `explain()` methods on the relation object to inspect its corresponding SQL query and the query plan.

Adopting a querying approach

As we've seen, the Relational API provides a range of flexible modes for composing DuckDB queries when working in Python. We've seen how we can define relations representing complex queries using SQL, which we can combine with replacement scans to compose pipelines of operations decoupled into multiple relations. We've also seen how we can use the Relational API as a dataframe-like query interface, which allows us to dynamically build up complex relations using method chaining and expression objects from the Expression API.

It's entirely up to you which approach you adopt, and you may find strategies are more appropriate in different situations. For people comfortable with SQL, they may find this their preferred query interface. The portions of the Relational API that enable dataframe-like usage are also very much still being developed, so there are times when you may run into feature gaps that mean you'll need to use SQL. As already mentioned, you can also combine these approaches; for example, by creating initial more complex relations using SQL, and then applying convenient method calls on the initial relation object as needed.

One of the primary benefits of leaning into the Relational API's dataframe-like mode of use is that it gives you tighter integration with Python language features, allowing you to interact with query components as first-class Python objects and dynamically construct them programmatically. This is particularly useful when developing DuckDB-powered products and services, where query logic may need to be customized at runtime in response to user input and other dynamic parameters. Another benefit is that, in contrast to defining monolithic queries as SQL strings, iteratively composing relations and creating expression objects will tend to provide more localized—and, therefore, more useful— query validation errors, providing a better development experience.

In its current form, the Relational API doesn't offer quite as streamlined a dataframe experience as libraries such as pandas or `dplyr` in R, though we look forward to enhancements in this direction. If you're particularly interested in using DuckDB in this mode, you might like to explore Ibis, an open source dataframe library that has support for using multiple data processing engines for its backend, including DuckDB. We will briefly explore using Ibis with DuckDB in *Chapter 12* when we cover alternative DuckDB query interfaces. You'll also find a discussion of DuckDB's Spark API, which enables querying DuckDB databases using PySpark, which also offers a dataframe-oriented interface.

We've covered a lot in this little side adventure around how to define queries and apply transformations with the Relational API; however, there's still more to cover. Next, we'll cover how to export data as files to disk using the Relational API.

Writing files to disk

The Relational API provides the following methods that allow you to write the query results of a relation object to CSV and Parquet:

- `DuckDBPyRelation.write_csv()`
- `DuckDBPyRelation.write_parquet()`

Let's use these methods to write our transformed pet dataset to disk as both CSV and Parquet files:

```
pets_cleaned_relation.write_csv("seattle_pets.csv")
pets_cleaned_relation.write_parquet("seattle_pets.parquet")
```

Just as we saw previously with the `read_csv()` and `read_parquet()` Relational API methods, `write_csv()` and `write_parquet()` both take a range of keyword arguments that correspond to parameters of DuckDB's CSV and Parquet writers. As always, you can inspect the available keyword arguments for these methods via Python's built-in `help()` function.

Again, as we saw earlier in the chapter when we were reading files, we can also use SQL to export data to files. This involves using the `COPY ... TO` statement:

```
duckdb.sql(
    "COPY pets_cleaned_relation TO 'seattle_pets.csv'"
)
duckdb.sql(
    "COPY pets_cleaned_relation TO 'seattle_pets.parquet'"
)
```

As we write this, there is no `DuckDBPyRelation.write_json()` Relational API method; however, you can export to JSON files using SQL:

```
duckdb.sql(
    "COPY pets_cleaned_relation TO 'seattle_pets.json'"
)
```

You can find the assorted parameters that can be used to configure exporting to different file formats in the DuckDB documentation for the `COPY` statement: `https://duckdb.org/docs/sql/statements/copy.html`.

Next, we'll look at how we can modify the contents of DuckDB databases using the Relational API.

Modifying the database

We can also use the Relational API to make modifications to databases. Given that we'll be making stateful modifications to some data, let's start by creating a new disk-based DuckDB database so that these changes can persist across Python processes:

```
conn = duckdb.connect("seattle_pets.db")
```

With this connection object, we can now start using the Relational API the same way as we were using the `duckdb` module before. The first thing we'll do is read the contents of the `seattle_pets.parquet` file we created earlier, which contains the contents of the transformed pets dataset. We'll then create a table in our new database using the `to_table()` method of the resulting relation, which will populate it with the contents of the Parquet file:

```
conn.read_parquet("seattle_pets.parquet").to_table("pets")
```

We now have a table called `pets` in our DuckDB database catalog that contains the contents of our transformed dataset. To confirm that we indeed successfully created the table, we can run a `SHOW TABLES` SQL statement on the database:

```
conn.sql("SHOW TABLES")
```

This creates a very simple relation with this output:

Figure 8.15 – A simple relation containing the name of each table in the current database catalog, which currently only contains the pets table

In order to interact with this table, we can create a relation object that references the contents of the pets table using the `table()` method:

```
conn.table("pets").count("*")
```

Again, this is another simple relation, with this output:

```
┌───────────────┐
│ count_star()  │
│     int64     │
├───────────────┤
│         42567 │
└───────────────┘
```

Figure 8.16 – A simple relation containing the number of records in the pets table

In the preceding example, we created a relation object that references the pets table, and then immediately called its `count()` method to see how many rows there are in the table.

Now that we've loaded a new database with our cleaned pets dataset, let's imagine that we've found some missing pet registration records that we need to update the pets table in the database with. One way we can insert a new record into this table is by using the `insert()` method of a relation that references a database table. To do this, we'll create a relation that references the pets table, and then we'll call the `insert()` method of the resulting relation, supplying a tuple containing the values of our new record:

```
new_dog1 = (
    "2024-05-19",
    "Monty",
    "Dog",
    "Border Collie",
)
conn.table("pets").insert(new_dog1)
```

Another way we can insert a new row into a table is by creating a new relation that contains a single record and then calling the `insert_into()` method on this relation, supplying it with the name of the table we want to insert the contents of the relation into:

```
new_dog2 = (
    "2024-05-19",
    "Pixie",
    "Dog",
    "Australian Kelpie",
)
```

```
new_dog_rel = conn.values(new_dog2)
new_dog_rel.insert_into("pets")
```

Going through the preceding code, we first created a tuple containing the values of our new record and then created a new relation from this using the `values()` method. We then called the `insert_into()` method of this new relation, passing the name of the table that this row should be inserted into.

To confirm that both of our new records have made it into the database, we'll query the `pets` table for entries with the `issue_date` we used:

```
conn.table("pets").filter("issue_date = '2024-05-19'")
```

This produces the following output:

issue_date date	pet_name varchar	species varchar	breed varchar
2024-05-19	Monty	Dog	Border Collie
2024-05-19	Pixie	Dog	Australian Kelpie

Figure 8.17 – A relation containing the two pet registration records that we added to the pets table

We can see that both new records have been returned, as expected. Note that if you are using a more recent version of the Seattle Pet Licenses dataset from the one included in our GitHub repository, and it includes registrations from May 2024, you may see more records than the two you just inserted.

Lastly, we'll close the database connection to ensure that all updates are written successfully to the database file:

```
conn.close()
```

This has concluded our whirlwind tour of DuckDB's Relational API for Python. While not a comprehensive guide, you should now have a sense of how to use it to build up lazily evaluated queries and inspect their results in the context of a data analysis workflow. For further reference, we suggest consulting the DuckDB documentation on the Relational API: https://duckdb.org/docs/api/python/relational_api.

In the next section, we'll move on to looking at the other way to interact with DuckDB databases in Python: the Python DB-API.

Working with the Python DB-API

The DuckDB Python client also provides another API for interacting with DuckDB databases: the Python DB-API. This is an interface that is compliant with the **Python Database API Specification v2.0 (DB-API 2.0)**, which is described by *PEP 249* (`https://peps.python.org/pep-0249`). The DB-API specification describes a standardized interface for accessing databases within Python, which has become a commonly used standard for Python database client packages. By encouraging conformity of API interfaces across different Python database libraries, Python code that targets this interface is more readily understood and more portable across databases, leading to a general increase in the availability of database connectivity in the Python ecosystem.

If you're developing a Python application that will potentially need to connect to a range of data sources, you will likely want to use the DB-API where possible, as the use of a standardized interface will enable more reusable code associated with database connectivity, resulting in a reduction in integration time for each new database connector being added. Another reason to consider using the DB-API interface is when your application code will be read by a diverse audience, such as in the context of an open source package. Using an interface that will be familiar to many developers will make for more accessible code.

In this section, we'll go on a quick tour of how to interact with DuckDB databases via the DB-API support, continuing to work with the Seattle Pet Licenses dataset from the previous section. Note that this is not intended to be an exhaustive treatment of DuckDB's DB-API. For that, we encourage you to consult the DuckDB documentation: `https://duckdb.org/docs/api/python/dbapi`.

Connecting to a database

As we mentioned in the previous chapter, when developing library packages that make connections to a DuckDB database, you should avoid using the default in-memory database and instead create and manage your own database connections so that consumers of your library can safely invoke DuckDB's default database for their own applications. These database connection objects, which we saw previously, and the `connect()` method that produces them are part of the DB-API specification.

Let's start by creating a new database connection to interact with:

```
conn = duckdb.connect()
```

This has created a new in-memory database. Please refer to *Chapter 7* for how to connect to other types of databases, such as disk-based databases.

Querying databases

In contrast to the Relational API, which provides a Python abstraction over SQL queries, the DB-API largely involves writing the SQL we want to run against the database directly. The primary way to send SQL to a target database is via the `execute()` method. Let's use this to create a table from the contents of our transformed Seattle pets dataset we saved to Parquet:

```
conn.execute(
    """
    CREATE OR REPLACE TABLE seattle_pets AS
    SELECT * FROM 'seattle_pets.parquet'
    """
)
```

This results in our new table being created immediately. You may also notice from the cell output that the `execute()` method returns the same connection object. Let's now submit a simple query that selects all results from this new `seattle_pets` table we just created:

```
conn.execute("SELECT * FROM seattle_pets")
```

Similar to when we were using DuckDB's Relational API, we don't get any results back from calling `execute()` immediately. To see the results of our query, we need to explicitly ask DuckDB for some results, which we do by calling methods against the connection object. Following the DB-API specification, the methods we have available from our connection object to do this are the following:

- `DuckDBPyConnection.fetchone()`
- `DuckDBPyConnection.fetchmany()`
- `DuckDBPyConnection.fetchall()`

As you might have guessed, the `fetchone()` method will retrieve a single row from the query result set:

```
conn.fetchone()
```

This returns a single row from the result set of the query as a tuple of values, with each value having the Python type that corresponds to that column's DuckDB type:

```
(datetime.date(2015, 12, 18), 'Zen', 'Cat', 'Domestic Longhair')
```

As the record is a simple tuple, it does not provide any context for the structure of the data being fetched. If needed, the `description` attribute of the connection object provides details about each column within the result set currently being fetched:

```
conn.description
```

This results in the following output:

```
[('issue_date', 'Date', None, None, None, None, None),
('pet_name', 'STRING', None, None, None, None, None),
('species', 'STRING', None, None, None, None, None),
('breed', 'STRING', None, None, None, None, None)]
```

Each tuple representing a column contains the column name and its type. Properties described by the DB-API 2.0 specification that DuckDB does not presently support are represented by `None`.

The connection object tracks the current position in the result set of rows that have been fetched, so you can continue to invoke `fetchone()` to get back the next result. The following list comprehension illustrates three successive calls to get the next row:

```
[conn.fetchone() for i in range(3)]
```

This results in the following output:

```
[(datetime.date(2016, 6, 14), 'Misty', 'Cat', 'Siberian'),
 (datetime.date(2016, 8, 4), 'Lyra', 'Cat', 'Mix'),
 (datetime.date(2019, 2, 13), 'Veronica', 'Cat', 'Domestic Longhair')]
```

A more direct way to achieve a list of the next multiple rows is to use the `fetchmany()` method:

```
conn.fetchmany(3)
```

This results in the following output showing us the next three rows:

```
[(datetime.date(2019, 8, 10), 'Spider', 'Cat', 'LaPerm'),
 (datetime.date(2019, 11, 21), 'Maxx', 'Cat', 'American Shorthair'),
 (datetime.date(2020, 5, 24), 'Mickey', 'Cat', 'Domestic Longhair')]
```

Finally, we can also retrieve a list of the remaining rows in the result set via the `fetchall()` method:

```
rest_rows = conn.fetchall()
len(rest_rows)
```

It gives us the length of the list of rows returned:

```
42560
```

Note that this will have returned the remaining rows after already fetching the first seven rows with the previous `fetchone()` and `fetchmany()` calls. If we had instead called `fetchall()` immediately after executing our SQL query, we would have fetched the entire result set at once.

Running SQL queries using prepared statements

DuckDB also provides support for prepared statements, which is a way to optimize SQL statements for reuse and which can accept parameters to customize each invocation of the prepared statement. This is a safer way to make your queries reusable within Python than using string interpolation (such as `f"WHERE species = '{species_val}'"` f-strings), which can be vulnerable to SQL injection exploitation.

To take advantage of prepared statements in the Python DuckDB client, we need to call the `execute()` method, passing in a `parameters` keyword argument that we can use to pass parameterized values into our SQL query.

Let's say we needed a repeatable way to add new pets to the database. There are a few ways we can construct prepared statements through parameterized variable values that can be specified programmatically. One way is through the use of `?` characters within the SQL query that indicate values to be filled via the `parameters` keyword argument, which needs to be provided with a list or tuple of values. In this case, we will be providing a list or tuple of values that represents a new pet record. The `?` variables will be filled in from left to right, in the same order they appear in the sequence supplied to the `parameters` keyword argument:

```python
import datetime
new_pet1 = (
    datetime.date.today(),
    "Ned",
    "Dog",
    "Border Collie",
)

conn.execute(
    "INSERT INTO seattle_pets VALUES (?, ?, ?, ?)",
    parameters=new_pet1
)
```

When the number of parameters starts to grow, this strategy can become hard to manage, and it also doesn't support reusing a single parameter across multiple variables. A more flexible approach to creating a prepared query is to use named parameters. This involves using named variables within your SQL string, which are represented by identifiers prefixed with the $ character, such as $breed. When using named parameters, the value of the parameters keyword argument must be a dictionary, whose keys and values correspond to the name and value of each parameter:

```python
new_pet2 = {
    "name": "Simon",
    "species": "Cat",
    "breed": "Bombay",
    "issue_date": datetime.date.today(),
}


conn.execute(
    """
    INSERT INTO seattle_pets VALUES
        ($issue_date, $name, $species, $breed)
    """,
    new_pet2
)
```

Lastly, let's double-check that our execute() call worked using the simple ? parameter substitution to pass today's date into an appropriate WHERE clause expression:

```python
conn.execute(
    """
    SELECT *
    FROM seattle_pets
    WHERE issue_date = ?;
    """,
    [datetime.date.today()],
).fetchall()
```

This gives us back the following two records we expected to see:

```
[(datetime.date(2023, 7, 25), 'Ned', 'Dog', 'Border Collie'),
 (datetime.date(2023, 7, 25), 'Simon', 'Cat', 'Bombay')]
```

Prepared statements provide a safe way to parameterize SQL queries without running the risk of opening up your code to SQL injection vulnerabilities. This is especially important when building and deploying a data application that accepts user input to be run against your database.

Writing to disk

To write CSV and Parquet files to disk, we just need to use the appropriate SQL statements and call them with `execute()`:

```
conn.execute(
    "COPY seattle_pets TO 'seattle_pets_updates.csv'"
)
conn.execute(
    "COPY seattle_pets TO 'seattle_pets_updates.parquet'"
)
```

This has now saved our transformed pets dataset along with the new rows to both CSV and Parquet files.

Closing the database connection

Finally, we'll make sure we close the database connection we have been working with.

```
conn.close()
```

Recall that in *Chapter 7*, we covered the importance of closing connections, as well as how to use connection objects as context managers via the `with` keyword to automatically close connections.

Database cursors

The DB-API also specifies that database cursor objects can be made using the `cursor()` method. A **cursor** is a mechanism provided by databases that enables programmatic traversal over query results. If you've worked with other Python DB-API libraries for different databases, you may have been surprised that we haven't been creating cursors to interact with the database. DuckDB does things a little differently, in that it puts the database interaction methods and properties on connection objects and does not have cursor objects as such.

In DuckDB's DB-API implementation, the `cursor()` method creates and returns a duplicate of the associated connection object. This is a feature that won't be needed all that frequently. One time you might want to create additional copies of a connection object is when you want to enable different threads to query the database in parallel. This is because connection objects themselves, while thread-safe, lock the database while the query is executing, preventing parallel execution. You can enable parallel execution by creating a new connection for each thread via the `cursor()` method on connection objects:

```
conn = duckdb.connect()
new_conn = conn.cursor()
```

In this example, we have created a new connection object, new_conn, which is a copy of the conn connection object. These two connection objects can now be used across threads in parallel. See the DuckDB documentation for an example of this in action: `https://duckdb.org/docs/guides/python/multiple_threads.html`.

We have just looked at two of the APIs that the Python client provides for interacting with and querying DuckDB: the Relational API and the DB-API. DuckDB's Python client also provides another API for interacting with DuckDB databases: the Spark API, which allows you to run PySpark queries against DuckDB databases. We won't cover this here; however, you can read more about the Spark API in *Chapter 12*.

Now that we've explored DuckDB's Relational API and its implementation of the DB-API, let's look at some features of the DuckDB Python client that are common to both.

Integration with Python packages and language features

This section outlines a range of integrations that the DuckDB Python client has across both Python language features and other Python packages that are commonly used in the Python data ecosystem. Note that all the integrations outlined here are applicable to both the Relational API and the DB-API.

Querying Python data structures

In addition to being able to query DuckDB relation objects, DuckDB is able to query directly from pandas Dataframes, Polars Dataframes, and Arrow tables. This means that you can treat objects of these types as if they were tables in your DuckDB database when constructing queries.

There are at least three distinct ways you can go about doing this. All of them work against pandas Dataframes, Polars Dataframes, and Arrow tables, as well as DuckDB relation objects. We'll demonstrate each of these three methods using a pandas dataframe that we'll create with pandas' read_parquet() function.

Querying Python objects via replacement scans

Replacement scans occur when a DuckDB query is made against an identifier that is not present in the current database's catalog. In this situation, alternative sources of table identifiers can be consulted, which for the Python client includes any objects able to be queried that are currently in scope. We already saw this in action with the Relational API, where we demonstrated writing a DuckDB query that queried the variable name of a DuckDBPyRelation object we had already created and that was in scope. DuckDB lets you query both types of Dataframes and Apache Arrow tables in the same way as well. Here's an example of querying our pandas dataframe:

```
import pandas as pd
pets_df = pd.read_parquet("seattle_pets.parquet")
duckdb.sql("SELECT * FROM pets_df USING SAMPLE 1")
```

This query retrieves a single row, randomly sampled from the pandas dataframe that we loaded from our Parquet file. Here's the row we retrieved (remembering that you'll likely sample a different row, due to our use of the SAMPLE clause):

```
┌────────────┬───────────┬─────────┬───────────────────────────┐
│ issue_date │ pet_name  │ species │           breed           │
│    date    │  varchar  │ varchar │          varchar          │
├────────────┼───────────┼─────────┼───────────────────────────┤
│ 2022-08-29 │ Maggie Mae│ Dog     │ Terrier, American Pit Bull│
└────────────┴───────────┴─────────┴───────────────────────────┘
```

Figure 8.18 – A row retrieved by querying a pandas dataframe using replacement scanning

Since the pets_df variable is currently in scope and refers to an object that DuckDB can consume, the query is successful via a replacement scan. This trick is particularly useful in the context of interactive data analysis workflows. In particular, when working within a notebook, you can conveniently query from pandas Dataframes, Polars Dataframes, and DuckDB relations that you have created in a previous cell.

Registering objects as DuckDB views

A limitation of using replacement scans to query other in-memory data structures is that their objects must be in scope. If your target dataframe object, for example, is located inside a dictionary as a value or is located within another module, then it would not be in scope and is thus not accessible in a query via a replacement scan.

To handle this limitation, you can register objects you want to be available to query in DuckDB's catalog as virtual tables, or views. These are analogous to SQL views, in that no new data is loaded into the database and the underlying data source is instead consulted each time a query is made.

Here's an example where our target pandas dataframe is stored inside the value of a dictionary, which means it cannot be queried via a replacement scan. Instead, we'll register it in the database's catalog with the name pets_view:

```
pets_dict = {
    "seattle": pd.read_parquet("seattle_pets.parquet")
}
duckdb.register("pets_view", pets_dict["seattle"])
duckdb.sql("SELECT * FROM pets_view USING SAMPLE 1")
```

After running this, we see a single row, randomly selected from the pandas dataframe that is stored inside our dictionary:

issue_date date	pet_name varchar	species varchar	breed varchar
2022-09-03	Samish	Dog	Retriever, Labrador

Figure 8.19 – A row retrieved by querying a pandas dataframe that
has been registered as a view in our DuckDB database

The preceding snippet illustrates that by registering our pandas dataframe in DuckDB's catalog as a view, we can query it within the FROM clause regardless of whether the corresponding object is in scope.

Creating tables from objects

Finally, in some circumstances, it may be more appropriate to create a new table with the contents of the target Python data structure, whether it be a pandas dataframe, a Polars dataframe, an Arrow table, or a DuckDB relation. Let's see this in action:

```
pets_df = pd.read_parquet("seattle_pets.parquet")
duckdb.sql(
    """
    CREATE OR REPLACE TABLE pets_table_from_df AS
    SELECT * FROM pets_df
    """
)
duckdb.sql("SELECT * FROM pets_table_from_df USING SAMPLE 1")
```

Once again, we see a single row, randomly selected from our pets dataset; however, this time, we've queried from a table that we loaded the contents of a pandas dataframe into:

issue_date date	pet_name varchar	species varchar	breed varchar
2022-06-25	Prim	Cat	Domestic Shorthair

Figure 8.20 – A row retrieved by querying a DuckDB table that we
loaded with the contents of a pandas dataframe

In the preceding snippet, we took our pandas dataframe containing a randomly shuffled instance of the pets dataset and created a table out of the records in the dataframe, using the column names and types from the dataframe to define the table's schema. Creating DuckDB tables from in-memory objects will incur overheads in having to copy the data from the dataframe; however, there are some circumstances, such as when needing to add new records to the dataset, in which you'll need a full DuckDB table to work with.

Now that we've covered the ways the DuckDB Python client allows you to readily consume Python data structures, let's switch to looking at how we can use it to convert query results to different Python data structures.

Converting query results

A notable contributor to the versatility of DuckDB's Python client is its ability to convert query results into a range of in-memory data formats, enabling it to integrate with a range of other tools in the Python data ecosystem.

We have already seen the `fetchone()`, `fetchmany()`, and `fetchall()` methods that retrieve records as either a single tuple or a list of tuples. While this rather generic format makes for a useful standardized data structure that improves interoperability, it is not overly efficient and is not well suited to integration with most analytical data tooling. Let's look at some of the other available data formats we can fetch results from. These are available against both relation objects and connection objects, which means that you can use the following methods regardless of whether you're using the Relational API or the DB-API:

- `df()`: Retrieves results as a pandas dataframe (pandas must be installed)
- `pl()`: Retrieves results as a Polars dataframe (Polars must be installed)
- `arrow()`: Retrieves results as an Arrow table (PyArrow must be installed)
- `fetchnumpy()`: Retrieves results as a dictionary of NumPy arrays (NumPy must be installed)
- `torch()`: Retrieves results as a dictionary of PyTorch Tensors (PyTorch must be installed)
- `tf()`: Retrieves results as a dictionary of TensorFlow Tensors (TensorFlow must be installed)

Note that this is not an exhaustive list; see the DuckDB documentation for reference (`https://duckdb.org/docs/api/python/result_conversion`). In this section, we'll go through examples of converting DuckDB results to pandas and Polars Dataframes, as well as Arrow tables.

Converting to dataframes

A common workflow when using DuckDB in analytical workflows involves querying a DuckDB database and then converting the results to a dataframe such as pandas or Polars, continuing to perform your analysis with the dataframe. This combines the strengths of both DuckDB and dataframe libraries, allowing you to crunch complex analytical queries over larger datasets on your local machine than most dataframe tools are able to handle, and then converting smaller result sets into a now manageable dataframe whose characteristic features lend themselves to effectively preparing and analyzing data. These characteristics include having more flexible schemas, inherent ordering of rows to ensure consistency of results during data analysis, and having APIs designed to facilitate rapid data wrangling. There is also a substantial amount of integration between dataframe libraries (especially pandas) and data visualization libraries. While not always necessary, converting results into a pandas dataframe will often be the easiest way to work with a data visualization library.

Over time, you will cultivate an intuition around typical points in your workflow where it makes sense to convert DuckDB results and start working with a dataframe. Right now, let's look at a very simple example that involves a hybrid DuckDB and pandas workflow. It involves loading our Seattle pets Parquet file into a DuckDB relation and converting its results into a pandas dataframe:

```
conn = duckdb.connect()
seattle_pets = conn.from_parquet("seattle_pets.parquet")

pandas_df = seattle_pets.df()
pandas_df[
    pandas_df["species"] == "Dog"
].value_counts("breed")[:5]
```

This results in a pandas `Series` object, which when displayed, produces the following output:

```
breed
Retriever, Labrador    3034
Retriever, Golden      1488
Chihuahua, Short Coat  1443
German Shepherd         954
Poodle, Miniature       846
Name: count, dtype: int64
```

Figure 8.21 – A pandas Series object containing the number of occurrences of each dog breed in the dataset

Once we've created a pandas dataframe from the relation object using the df () method, we can jump into using pandas to analyze the top five most popular breeds of dogs for pets in Seattle. Let's now illustrate the conversion to a Polars dataframe by using it to analyze the top five cat breeds:

```python
import polars as pl

polars_df = seattle_pets.pl()
polars_df.filter(
    pl.col("species") == "Cat"
)["breed"].value_counts(sort=True)[:5]
```

This results in a Polars DataFrame object, which when displayed, produces the following output:

shape: (5, 2)

breed	count
str	u32
"Domestic Short...	7408
"Domestic Mediu...	1555
"American Short...	1301
"Domestic Longh...	881
"Siamese"	457

Figure 8.22 – A Polars DataFrame object containing the number
of occurrences of each cat breed in the dataset

Just as with pandas, we can see that DuckDB makes it straightforward to convert the results of our query to a Polars dataframe and continue to work with our data as a dataframe.

The logic we applied using dataframes in both these examples was pretty simple, and we could have just as easily used DuckDB to perform the complete operation. These examples are just intended to illustrate the mechanics of workflows that combine both DuckDB and dataframes rather than being motivated by the need for a dataframe library. In *Chapter 11*, we'll see an applied workflow that makes use of a hybrid DuckDB and pandas workflow to efficiently crunch a large dataset while also gaining the benefits of pandas integration for data visualization.

Converting to Arrow tables

Apache Arrow is a particularly useful in-memory data format that DuckDB can convert results into. A notable benefit in converting to Arrow tables is that this will be a zero-copy operation from the DuckDB database, which means the desired data does not need to be copied to a new memory location for the new data structure, saving both memory bandwidth and CPU cycles. Converting to other data formats will involve some degree of conversion overheads, which you may need to consider for certain applications.

Arrow tables are also useful as a data interchange format, as many popular in-memory data format libraries can consume Arrow tables. This makes it useful for converting to other data formats, such as Vaex and Apache Arrow DataFusion, which DuckDB does not support converting to directly at the time of writing.

Here, we see an example of how to convert DuckDB query results to an Arrow table, this time using the DB-API's `execute()` method on a new database connection object:

```
conn = duckdb.connect()
conn.execute("SELECT * FROM 'seattle_pets.parquet'")
pets_table = conn.arrow()
pets_table.schema
```

Evaluating the schema attribute of the `pets_table` Arrow table produces the following output:

```
issue_date: date32[day]
pet_name: string
species: string
breed: string
```

As you can see, when using the DB-API, we need to first submit the query to the database using the `execute()` method. Then, we call the desired result conversion method (in this case, `arrow()`) to instruct the database to perform the query and return the result converted into the requested format. Alternatively, we could have used the `sql()` method from the Relational API instead of the `connect()` method. Doing this with our preceding example would have instead created an intermediate relation object and then converted its result set to an Arrow table.

Data types – from Python to DuckDB

As we have discussed previously, DuckDB has a comprehensive range of data types, which are outlined in the DuckDB documentation: `https://duckdb.org/docs/sql/data_types/overview`. When working with the Python DuckDB client, one consideration is how Python types are mapped to DuckDB types. Within the Python API, DuckDB types are represented by instances of the `DuckDBPyType` class. You can access these instances directly via the `duckdb.typing` module:

```
varchar_type = duckdb.typing.VARCHAR
bigint_type = duckdb.typing.BIGINT
```

In general, any method in the DuckDB Python API that accepts a `DuckDBPyType` instance can be given a Python type, and it will be automatically converted according to the mapping found in the DuckDB Python client documentation: `https://duckdb.org/docs/api/python/types`. You can also manually create instances of `DuckDBPyType` by passing in the desired Python type, which will be mapped to the corresponding DuckDB data type:

```
varchar_type = duckdb.typing.DuckDBPyType(str)
bigint_type = duckdb.typing.DuckDBPyType(int)
```

Python values are also coerced into appropriate DuckDB values where possible. To see this in action, we can use the `values()` method from the Relational API, which takes a sequence of values and returns a DuckDB relation representing these values as a single row with appropriately typed columns. Here's an example with a range of Python data types, illustrating the automatic conversion:

```
duckdb.values(
    [
        10,
        1_000_000,
        0.95,
        "hello string",
        b"hello bytes",
        True,
        datetime.date.today(),
        None,
    ]
)
```

This produces a relation that is displayed like this:

col0 int32	col1 int32	col2 double	col3 varchar	col4 blob	col5 boolean	col6 date	col7 int32
10	1000000	0.95	hello string	hello bytes	true	2024-05-19	NULL

Figure 8.23 – A relation containing a single record of values created using the values() method

We can see, for example, that the Python integer values have been coerced into appropriately sized DuckDB integer types, the string into a VARCHAR type, and Python datetime.date into a DuckDB DATE type.

Also supported is the conversion of nested data types, such as lists, tuples, and dictionaries:

```
duckdb.values(
    [
        (1, 2),
        ["hello", "world"],
        {"key1": 10, "key2": "quack!"}
    ]
)
```

This produces a relation that is displayed like this:

col0 int8[]	col1 varchar[]	col2 struct(key1 tinyint, key2 varchar)
[1, 2]	[hello, world]	{'key1': 10, 'key2': quack!}

Figure 8.24 – A relation containing a single record of nested values created using the values() method

We can see that both the tuple and the list values have been converted to DuckDB LIST types, which are ordered sequences all of the same type. We can also see that the dictionary has been converted into a DuckDB STRUCT type; structs are mappings between string keys and values, whose type can vary across each key.

This has provided a quick tour of how to think about the interface between Python and DuckDB types. For more details, see the DuckDB documentation for working with Python types (https://duckdb.org/docs/api/python/types) and the DuckDB data types reference (https://duckdb.org/docs/sql/data_types/overview).

In the next section, we'll look at registering Python functions for use within DuckDB, which is one place where we'll need to consider the mapping between Python and DuckDB types.

User-defined functions

There are times when you might find yourself wanting some capabilities provided by Python or its ecosystem of packages when writing DuckDB SQL queries. For such situations, DuckDB offers the ability to create **user-defined functions** (UDFs). This enables you to register Python-defined functions within a DuckDB database and use them as though they were DuckDB SQL functions.

To illustrate this feature, let's imagine that we wanted to augment our pets dataset with a column containing an emoji corresponding to the species of each pet's registration record. We'll start by defining a Python function, emojify(), that makes use of the Python emoji package (https://github.com/carpedm20/emoji) to convert a species name into a string containing the corresponding emoji. Note that the emoji package will have been installed when installing the contents of requirements.txt in the *Technical requirements* section of *Chapter 7*. You can also install it directly by running pip install emoji on the command line:

```
import emoji
def emojify(species):
    """Converts a string into a single emoji."""
    emoji_str = emoji.emojize(f":{species.lower()}:")
    if emoji.is_emoji(emoji_str):
        return emoji_str
    return None
```

This Python function takes a string and attempts to convert it into a Unicode emoji, returning None if it cannot. Let's try it out:

```
emojify("goat")
```

As hoped, we get a string containing a goat emoji as our output:

```
'🐐'
```

We want to be able to apply this to every pet record, so the next thing we need to do is register this function with DuckDB. To do this, we need to use the create_function() method. In this case, we're registering it on the default database, but we could also be calling it against a specific connection object. To register a function, we need to specify the following information as arguments to the create_function() method:

- The function name will be registered as in DuckDB; we'll name it the same as the Python function name.
- The Python function object to register.
- A list of DuckDB types corresponding to the type of each function argument. In this case, it will be a single VARCHAR type since our Python function takes a single argument of type str.
- The return type of the function, which will also be VARCHAR.

Taking the aforementioned into account, our `create_function()` call will look like this:

```
duckdb.create_function(
    "emojify",
    emojify,
    [duckdb.typing.VARCHAR],
    duckdb.typing.VARCHAR
)
```

With the `emoji` function registered, we can now use it in a SQL query as if it were a DuckDB SQL function. Let's query the Parquet file we saved with updated additional pet records, returning all columns as well as the newly derived `emoji` column that contains the emoji representation of each pet. We'll also use the `SAMPLE` clause so that we get a random sample of 10 pet records to review:

```
duckdb.sql(
    """
    SELECT *, emojify(species) AS emoji
    FROM 'seattle_pets_updates.parquet'
    USING SAMPLE 10
    """
)
```

This results in the following output:

issue_date date	pet_name varchar	species varchar	breed varchar	emoji varchar
2022-04-18	Luella	Dog	Havanese	🐕
2022-02-25	Bo-Riley Oh	Dog	Poodle, Miniature	🐕
2022-12-24	Sammy	Dog	Anatolian Shepherd	🐕
2021-11-11	Mayer	Dog	Retriever, Golden	🐕
2023-01-16	Arnie	Dog	Retriever	🐕
2022-05-14	Benny	Dog	Terrier	🐕
2022-10-19	Bommeke	Cat	Domestic Longhair	🐈
2022-12-10	Poppy	Dog	Spaniel, Welsh Springer	🐕
2021-04-10	Katniss	Cat	Mix	🐈
2021-08-18	Maxwell	Dog	Pointer, German Shorthaired	🐕
10 rows				5 columns

Figure 8.25 – A relation with 10 values sampled from the cleaned pet registrations dataset, which have been augmented with an emoji corresponding to the type of animal

Success! This illustrates how once we have defined and registered a Python function, we can readily use it within a SQL query, just as if it were a DuckDB SQL function.

In order to remove the UDF from the database, we just call the `remove_function()` method:

```
duckdb.remove_function("emojify")
```

In the previous section on Python and DuckDB typing, we mentioned that anywhere a DuckDB type is required in a DuckDB Python API method, the corresponding Python type can be used instead, and it will be automatically converted. Let's see that in action with the `create_function()` method:

```
duckdb.create_function("emojify", emojify, [str], str)
```

We can simplify the registration of this function further by defining our Python `emojify()` function using type annotations on both the function arguments and the return value. DuckDB can inspect these annotations and automatically infer the correct types for registering the function:

```
def emojify(species: str) -> str:
    """Converts a string into a single emoji."""
    emoji_str = emoji.emojize(f":{species.lower()}:")
    if emoji.is_emoji(emoji_str):
        return emoji_str
    return None

duckdb.remove_function("emojify")
duckdb.create_function("emojify", emojify)
```

With our Python function redefined to make use of type annotations for its input and output types, we can now omit those types when registering our UDF, only requiring the name of the function to create and the Python function object itself.

Performance considerations for Python UDFs

An important consideration around the use of Python UDFs is that they will introduce a performance hit when compared with using native DuckDB functions. When a UDF is called in a query, DuckDB will process chunks of data individually by threading each one through the local Python interpreter, which will slow your query down. DuckDB does support the creation of more efficient vectorized UDFs that accept Arrow arrays as input; however, there will still be a performance penalty incurred. Where you have the option, you should therefore use native DuckDB functions, reaching for UDFs only when DuckDB doesn't provide your required functionality.

DuckDB's UDF support offers a few more features that we won't go into here, such as creating more efficient vectorized UDFs that accept Arrow arrays as input, enabling functions that have side effects, and controlling how the function behaves when NULL values are passed as inputs. We refer you to the DuckDB documentation for a more complete treatment of its UDF support: `https://duckdb.org/docs/api/python/function`.

Handling exceptions

When writing application code, it's good practice to make your code resilient to runtime exceptions through the use of error handling. The DuckDB Python client has a collection of exception classes that it uses for raising appropriate exceptions given the error that has occurred. Following exception-handling best practice, you should, in general, aim to catch the most specific exception type that could arise within the line or lines of code you are guarding. The DuckDB exception classes are located at the top level of the `duckdb` module. Here, we'll just briefly look at a couple of available exceptions and situations where they might be needed, to give you a sense of how to work with them.

Let's start with how you can handle errors that arise when DuckDB cannot convert a value to a target type:

```
from duckdb import ConversionException
try:
    duckdb.execute("SELECT '5,000'::INTEGER").fetchall()
except ConversionException as error:
    print(error)
    # handle exception...
```

The error that we caught and printed results in this output:

```
Conversion Error: Could not convert string '5,000' to INT32
```

The preceding snippet catches the DuckDB `ConversionException` exception, which was thrown due to a conversion from string to integer failing on account of a non-digit character occurring in the string. For the next type of error, let's look at how you can guard against DuckDB making a query against a non-existent table in the database catalog:

```
from duckdb import CatalogException
try:
    duckdb.sql("SELECT * from imaginary_table")
except CatalogException as error:
    print(error)
    # handle exception...
```

The error that we caught and printed results in this output:

```
Catalog Error: Table with name imaginary_table does not exist!
Did you mean "pg_tables"?
```

This snippet catches the DuckDB `CatalogException` exception, which was caused by querying an identifier that does not exist in the database catalog as a table or other source. You're then able to put whatever appropriate handling logic is needed to account for this scenario within the `except` block.

For a complete list of available exceptions provided by the DuckDB Python client, see the full DuckDB Python client API reference: `https://duckdb.org/docs/api/python/reference`.

Summary

In this chapter, we went on a deep dive through the DuckDB Python client. We started with the Relational API, illustrating its affinity for analytical workflows by using it to explore the Seattle pet licenses dataset. We then took the DB-API through its paces, illustrating some of the operations that you can use to build data applications and integrations. We then finished the chapter with a look at a range of Python integrations that DuckDB offers, including working with other data structures, converting results into other formats, type conversion between Python and DuckDB, creating UDFs, and lastly, handling exceptions raised by DuckDB. Do note that this chapter and the previous one are not intended to be an exhaustive guide to the DuckDB Python API. For that, you should consult the documentation for the DuckDB Python API: `https://duckdb.org/docs/api/python`.

You now have the ingredients you need to bring DuckDB into your Python data analysis workflows or to start building data applications or integrations in Python using DuckDB. After going through the Python DuckDB client in some depth, in the next chapter, we will jump into using the DuckDB client for R, a language designed specifically to support statistical computing applications and that is notable for its popularity among scientists, data scientists, and data analysts.

9

Exploring DuckDB's R API

In this chapter, we'll be taking a close look at another DuckDB client API; this time for the R programming language. R is a programming language that's designed for statistical computing and graphically representing data. It is used by many scientists and data practitioners, including data scientists, statisticians, and data analysts. It's well-suited for performing data analysis, statistical modeling, and data visualization, as well as developing statistical software. DuckDB provides a feature-rich package for working with DuckDB databases in R: https://r.duckdb.org.

R benefits from a rich array of built-in data processing features and a rich ecosystem of data libraries. Experienced R users may wonder why they might want to add another R package to their toolbox. DuckDB enables you to scale your analytical data workflows further – you can run complex analytical operations efficiently and process much larger datasets than would be otherwise possible on a local machine. Furthermore, as we have seen, DuckDB is not just a data processing tool. As a fully-fledged **database management system (DBMS)**, DuckDB enables you to uplift your analytical workflows in R with data management features not available in most data processing tools. Its versatile feature set allows it to both read and write to a wide variety of data sources, including local and remote-hosted file formats, run queries over data in external databases, and apply efficient querying patterns against data lakes. Rather than necessarily replacing existing tools, DuckDB complements R features and packages. DuckDB can both query and export to R dataframes; you can connect to DuckDB databases and interact with them via R's **Database Interface (DBI)**; you can query DuckDB databases with dplyr queries; and DuckDB also makes for a perfect low-latency analytical query engine for powering Shiny data apps.

In this chapter, we'll cover the following aspects of working with DuckDB and R:

- Setting up your environment to work with R and DuckDB
- Connecting to and working with DuckDB using R's DBI package
- Registering and querying R objects as virtual tables
- Querying DuckDB using the dplyr package

By the end of this chapter, you'll be all set to bring DuckDB into your R data analysis workflows and use it alongside your existing tools to scale and streamline your analytical workflows.

Technical requirements

Before we jump into using the DuckDB R client, we need to make sure that we have an appropriate environment up and running. In this section, we'll go through the following aspects:

- Installing R (4.1 or later) on your machine
- Installing an IDE that supports R and offers a notebook interface
- Installing the R packages that we'll need to work with DuckDB

The R code for this chapter can be found in the `chapter_09` directory of this book's GitHub repository: `https://github.com/PacktPublishing/Getting-Started-with-DuckDB/tree/main/chapter_09`.

> **Getting the most out of this chapter**
>
> We've tried to make this chapter as accessible as possible to newcomers to R, but it's not a complete introduction to R or doing data analysis with R. You'll get the most out of this chapter if you've had some hands-on experience with R, including working with dataframes, however you can also use it as a brief crash course in getting started with R for data analysis. If you're interested in diving deeper into learning about R for data analysis, we recommend another Packt book, *Data Wrangling with R*.

Installing R

The first step is to make sure you have R installed on your machine. The exercises we'll go through require at least R 4.1.0. If you do not already have a working installation of R 4.1.0 or later, go to the R project's web page at `https://cloud.r-project.org` and follow the appropriate instructions for your operating system on how to download and then install R.

Getting your IDE set up

The examples in this chapter involve sequential steps when analyzing a dataset, so they are best worked through in a notebook environment. There's a range of IDEs with notebook support that would be suitable to use for the R examples we'll cover. R users may prefer to use RStudio to work with R Markdown, and those more familiar with Python may prefer to use the Jupyter Notebooks via JupyterLab or VS Code. We'll briefly go through each option here, providing suggestions for which you may want to use and directions for how to get each installed.

RStudio

The most popular notebook format in the R ecosystem is R Markdown, which people commonly use RStudio to work with. If you are already familiar with using RStudio for working with R Markdown documents, or you happen to already have RStudio installed, you may want to consider this option.

As you already installed R in the previous section, the only thing you need to do is download and install RStudio itself by following the instructions on the home page for Posit (`https://posit.co`), the developers of RStudio.

Alternatively, you can use VS Code to work with R Markdown. See VS Code's documentation for how to set up this integration: `https://code.visualstudio.com/docs/languages/r`.

Jupyter Notebooks

While the Jupyter Notebook platform is commonly associated with notebooks that run Python, it also supports notebooks that run R. If the Jupyter Notebook interface is more familiar to you, whether from previous experience or just from working with it in the previous chapter, then you may want to consider this option.

The simplest way to use R in Jupyter Notebooks is by using **IRKernel**, which is a native R kernel for Jupyter. IRKernel's only Python dependency is that it requires access to the `jupyter` command of the target Jupyter installation. You should have this installed in your Python virtual environment already since you used it to complete the exercises provided in *Chapter 7* and *Chapter 8*. If you don't have this set up, go to the *Technical requirements* section of *Chapter 7* and complete those instructions. To install IRKernel, go to its home page and follow the installation instructions (`https://irkernel.github.io`), remembering to activate your Python virtual environment first so that the `jupyter` command is available.

To get started working with R in JupyterLab, once you've launched JupyterLab, when you go to create a new notebook, you should have the option to create an R kernel. You can also change the kernel that's associated with any open notebook file by clicking on the kernel type in the top-right corner of the pane containing your notebook.

If you're more comfortable working with VS Code, instead of using JupyterLab, you can leverage VS Code's Jupyter Notebook support. This requires you to follow the same instructions for installing IRKernel, but then you'll use VS Code rather than JupyterLab to work with your notebook. To do this, after opening VS Code, create a new notebook by pressing *Ctrl + Shift + P* to open the command palette, and then run the `Create: New Jupyter Notebook` command. In the notebook that's created, you need to select the active kernel for this notebook in the top-right corner. Here, you want to select the one simply called *R*.

Installing the R dependencies

Finally, we'll install the R packages that are needed to run the examples in this chapter. Open the IDE that you've set up for working with R-flavored notebooks, create a new code cell, and enter the following R code:

```
install.packages(c("duckdb", "tidyverse", "arrow"))
```

Once you've run this cell, these R packages will be installed into your R environment. Note that if you're using a Linux operating system, this may take some time.

Getting the dataset

The data and example code for this chapter can be found in the `chapter_09` directory of this book's GitHub repository: `https://github.com/PacktPublishing/Getting-Started-with-DuckDB/tree/main/chapter_09`.

For this chapter, we'll be working with the NYC Dog Licensing Dataset, which contains records of dog licenses issued to residents of the city of New York. The dataset is made available through NYC Open Data: `https://opendata.cityofnewyork.us`. We have included a snapshot of this dataset in this chapter's directory, in the file that was used in this chapter, made available as the `NYC_Dog_Licensing_Dataset.csv` file.

Alternatively, you can download the most recently updated dataset from NYC Open Data: `https://data.cityofnewyork.us/Health/NYC-Dog-Licensing-Dataset/nu7n-tubp`. The file that was used to produce the output shown in this chapter was `NYC_Dog_Licensing_Dataset_20240429.csv`, which corresponds to the current edition of this dataset at the time of writing. We have renamed this `NYC_Dog_Licensing_Dataset.csv`. Note that you'll need to update the CSV-reading queries in this chapter so that they match the name of your downloaded CSV. It's also worth bearing in mind that more recent versions of the dataset will include subsequent dog licenses that have been issued. This means that your results might look a little different from those we have provided.

Opening the notebook

We're now ready to open the notebook containing the R examples for this chapter. We've provided two versions of the notebook in this book's GitHub repository: an R Markdown file for RStudio and a Jupyter Notebook file for use with JupyterLab (or other Jupyter-Notebook-compatible IDEs).

With your chosen IDE, open the relevant file for you:

- **RStudio**: `chapter_09.Rmd`
- **JupyterLab**: `chapter_09.ipynb`

Now that we've got our R environment and notebook IDE set up, we're ready to start working with DuckDB with R. In the next section, we'll get started by looking at how to connect to DuckDB databases.

Working with DuckDB using R's DBI

The DuckDB R client provides excellent support for the R DBI (https://dbi.r-dbi.org), a standardized API for communicating with DBMS. It's the standard way of connecting to and querying from databases in R. It is similar in design to the Python DB-API, which we discussed in *Chapter 8*, with the DBI package aiming to provide a single standardized database interface across the R ecosystem. This enables both R users and R package developers to work with and build against a single interface, rather than having to target a separate API for each target database.

The DBI package contains an extensive range of functions; you can find the complete reference at https://dbi.r-dbi.org/reference. In this section, we'll cover a subset of the DBI functions that you are likely to find useful for working with DuckDB in R. The core functions that you will want to be comfortable with are as follows:

- `DBI::dbConnect()`: Create a new connection to a target database

- `DBI::dbGetQuery()`: Run a SQL query that retrieves a result set and returns those results

- `DBI::dbExecute()`: Execute a SQL command that does not retrieve results against a database connection

- `DBI::dbDisconnect()`: Disconnect an existing connection object to a database

For these functions to be available in your session, we need to load the DBI package:

```
library("DBI")
```

With the DBI package loaded, let's get stuck into using it to work with DuckDB.

Connecting to DuckDB

To interact with DuckDB in R, we need to create a connection to a DuckDB database using the `DBI::dbConnect()` function. The first argument for this function is a driver object for the database you want to connect to. We can create a DuckDB driver object using the `duckdb::duckdb()` expression. This will be constant across all connections we make when working with DuckDB in R. The `dbdir` argument describes the target DuckDB database.

Now, let's run through how to create different types of DuckDB database connections. We'll start by creating a connection to an on-disk DuckDB database with the `quack.duckdb` file path:

```
disk_conn <- dbConnect(
    duckdb::duckdb(),
    dbdir = "quack.duckdb"
)
```

As this database didn't already exist, this will result in a connection to a new on-disk DuckDB database. If this path existed and contained a DuckDB database, it would have resulted in a connection to the existing database. In either case, we now have a DuckDB connection object that we can use to make persistent state changes. Note that the database file will be created in the current working directory of your R session. You can check this location by running the `getwd()` function. In situations where multiple processes need to access a single on-disk database concurrently, you must ensure that each client is connected in *read-only mode*. Read-only connections to on-disk databases can be enabled via the `read_only` argument, which takes a Boolean value:

```
read_only_conn <- dbConnect(
    duckdb::duckdb(),
    dbdir = "quack.duckdb",
    read_only = TRUE
)
```

Instead of creating an on-disk database, we can also create an in-memory database, which we can use when we don't need to persist a DuckDB database to disk. To do this, pass the special `":memory:"` string to the `dbdir` argument:

```
mem_conn <- dbConnect(
    duckdb::duckdb(),
    dbdir = ":memory:"
)
```

Now that we've established how to create different types of DuckDB database connections, including how to create new disk-based persistent databases, we'll continue working with DuckDB through the DBI package. Note that we'll be creating new connections as we continue, rather than using the various ones we just created.

Reading and writing tables

We'll start by looking at how we can write data from R into our DuckDB database. One context this can occur is when you have a dataframe that you want to write to a table in your DuckDB database. To motivate this situation, let's start by creating a dataframe by reading the NYC Dog License dataset from its CSV file.

We'll be using a function from the `tidyverse` ecosystem, so let's start by loading it. Note that you might see some warnings about conflicting names imported from the package; you can safely ignore this:

```
library(tidyverse)
```

We'll use the `readr::read_csv()` function from `tidyverse`, which allows us to read a CSV file into a dataframe, as well as provide a way to specify how specific columns should be parsed into R data types. One notable thing about data-returning functions from tidyverse is that rather than returning base R dataframes, which are of the `data.frame` type. they return *tibbles*. These are the tidyverse's leaner and more efficient dataframes and can be used interchangeably with standard R dataframes; in addition to `tbl_df`, they also inherit from `data.frame`.

Here's the code that will load our CSV file into a tibble:

```
dogs_df <- read_csv(
    file = "NYC_Dog_Licensing_Dataset.csv",
    col_types = cols(
        LicenseIssuedDate = col_date("%m/%d/%Y"),
        LicenseExpiredDate = col_date("%m/%d/%Y"),
        AnimalGender = col_factor(levels = c("M", "F")),
    ),
)
head(dogs_df, 5)
```

You may see a warning about some records from our CSV failing to parse correctly due to data quality issues in the dataset. This won't be an issue, but it's worth noting as we'll learn how we can handle this when we start using DuckDB later. Since we ended this example by passing our new dataframe into the `utils::head()` function, the output you'll see on running this code is a dataframe containing the first five rows of our ingested dataset:

AnimalName	AnimalGender	AnimalBirthYear	BreedName	ZipCode	LicenseIssuedDate	LicenseExpiredDate	Extract Year
<chr>	<fct>	<dbl>	<chr>	<dbl>	<date>	<date>	<dbl>
PAIGE	F	2014	American Pit Bull Mix / Pit Bull Mix	10035	2014-09-12	2017-09-12	2016
YOGI	M	2010	Boxer	10465	2014-09-12	2017-10-02	2016
ALI	M	2014	Basenji	10013	2014-09-12	2019-09-12	2016
QUEEN	F	2013	Akita Crossbreed	10013	2014-09-12	2017-09-12	2016
LOLA	F	2009	Maltese	10028	2014-09-12	2017-10-09	2016

A tibble: 5 × 8

Figure 9.1 – The first five rows of the dataframe derived from the NYC dog registrations CSV file

To read this CSV into a tibble, we used the `col_types` argument to specify the type and format of the `LicenseIssuedDate` and `LicenseExpiredDate` columns, whose non-standard date format can't be detected automatically by the CSV reader. We also specified that the `AnimalGender` field should be parsed as a **factor**, which is how R represents categorical values, with its two possible string values being M and F.

Now that we have a tibble containing the loaded dataset, we want to write it to a table in a new DuckDB database. Let's create a connection to a new on-disk DuckDB database, which we'll name `nyc_dogs.duckdb`:

```
conn <- dbConnect(
    duckdb::duckdb(),
    dbdir = "nyc_dogs.duckdb"
)
```

Now, we'll write the contents of our tibble to our database using the `DBI::dbWriteTable()` function. This function takes the name of the target table we wish to create and either a dataframe or a tibble whose contents we want to write to that table:

```
dbWriteTable(
    conn,
    "dogs_table_from_df",
    dogs_df,
    overwrite = TRUE
)
```

With this executed, we have created a table called `dogs_table_from_df` containing the contents of our `dogs_df` tibble. We can confirm this by using `DBI::dbListTables()` to inspect the contents of our DuckDB database's catalog:

```
dbListTables(conn)
```

We get the result as follows:

```
'dogs_table_from_df'
```

This tells us that our database's catalog contains the single table, `dogs_from_df`, which we just created. We can also use the `DBI::dbListFields()` function to inspect the column names for the table we just created:

```
dbListFields(conn, "dogs_table_from_df")
```

We get the following result:

```
'AnimalName'·'AnimalGender'·'AnimalBirthYear'·'
BreedName''ZipCode'·'LicenseIssuedDate'·'LicenseExpiredDate'·'Extract
Year'
```

Another frequently occurring situation is the reverse of what we just did: needing to convert a DuckDB table into a dataframe that we can work with in R. You can do this by using the `DBI::dbReadTable()` function. Let's do this now by converting our `dogs_table_from_df` table back into a dataframe. We'll inspect the final five rows of it using the `utils::tail()` function:

```
dogs_from_duckdb_df <- dbReadTable(
    conn,
    "dogs_table_from_df"
)
tail(dogs_from_duckdb_df, 5)
```

This gives us a dataframe containing the last five rows in the table:

	AnimalName	AnimalGender	AnimalBirthYear	BreedName	ZipCode	LicenseIssuedDate	LicenseExpiredDate	Extract.Year
	<chr>	<fct>	<dbl>	<chr>	<dbl>	<date>	<date>	<dbl>
616886	SKYE	F	2016	Great Pyrenees	11218	2023-11-01	2024-12-02	2023
616887	UNKNOWN	F	2023	Shih Tzu Crossbreed	10022	2023-11-01	2024-11-01	2023
616888	MUNYU	M	2009	Poodle, Toy	11355	2023-11-01	2024-11-24	2023
616889	SAINT	M	2021	Unknown	11412	2023-11-01	2024-11-01	2023
616890	BABY	F	2021	Unknown	10473	2023-11-01	2024-11-01	2023

A data.frame: 5 × 8

Figure 9.2 – The last five rows of the dataframe derived from the dogs_table_from_df DuckDB table

As you can see from the output, `DBI::dbReadTable()` returns a dataframe rather than a tibble. If you need a tibble instead, use the tidyverse's `tibble::as_tibble()` to convert the returned dataframe. In this case, our new dataframe simply contains the same data as the tibble we used to create the table, with the goal being just to illustrate a round trip conversion from R into a DuckDB table and back again.

Another way we can create a table in DuckDB using the DBI package is with the `DBI::dbCreateTable()` function, which will create a new table with a desired name and schema, but without any data. We can call this function in two ways. The first is by passing a dataframe or tibble into the third argument, which will cause DuckDB to create a new table using the column names and data types from the dataframe to create the table's schema. However, no records from the dataframe will be written:

```
dbCreateTable(conn, "empty_dogs_table", dogs_df)
dbReadTable(conn, "empty_dogs_table")
```

Note that if you've already run this cell previously, you'll need to call `dbRemoveTable()` on the table to remove it first. Running this code gives us the following empty dataframe:

AnimalName	AnimalGender	AnimalBirthYear	BreedName	ZipCode	LicenseIssuedDate	LicenseExpiredDate	Extract.Year
<chr>	<chr>	<dbl>	<chr>	<dbl>	<date>	<date>	<dbl>

A data.frame: 0 × 8

Figure 9.3 – A dataframe showing that empty_dogs_table has been
created with the desired schema but contains no rows

As we can see from retrieving our new table with `DBI::dbReadTable()`, the new `empty_dogs_table` table was successfully created with the schema from the input tibble and has no records.

The other way we can use `DBI::dbCreateTable()` to create a new table is by specifying the column names and types directly. Rather than passing in a dataframe or tibble into the third argument, we can pass in a named character vector whose element's names and values indicate the column names and R types, respectively:

```
dbCreateTable(
    conn,
    "good_dogs_table",
    c(name = "character", birthday = "date", age = "double")
)
dbListTables(conn)
```

We get the result as follows:

```
'dogs_table_from_df' · 'empty_dogs_table' · 'good_dogs_table'
```

As we can see from calling `DBI::dbListTables()`, the new table, `good_dogs_table`, has been created successfully. Note that the target data types specified in the character vector are string representations of R types. These will then be converted automatically into the appropriate DuckDB data types. You can confirm how R types will be mapped into DuckDB data types using the `DBI::dbDataType()` function, which takes R objects and returns strings representing SQL data types:

```
dbDataType(conn, "Monty")
dbDataType(conn, 6)
dbDataType(conn, today())
```

The output is as follows:

```
'STRING'
'DOUBLE'
'DATE'
```

Finally, we can also delete tables using the DBI::dbRemoveTable() function:

```
dbRemoveTable(conn, "good_dogs_table")
dbListTables(conn)
```

We get the following result:

```
'dogs_table_from_df'·'empty_dogs_table'We can see that we've
successfully removed the good_dogs_table table from our DuckDB
database.
```

We've only covered a handful of the DBI functions that you can use to read and write to DuckDB tables. To explore additional functions, consult the DBI API documentation: https://dbi.r-dbi.org/reference/#tables.

Next, we'll move on to using the DBI API to both query and execute SQL statements against DuckDB databases.

Querying and executing SQL statements

Let's move on to using the DBI package to query DuckDB databases with SQL and also to execute other types of SQL statements. The DBI::dbGetQuery() and DBI::dbExecute() functions take SQL strings as arguments, with the difference being that you will use DBI::dbGetQuery() for SQL queries that return result sets, whereas the DBI::dbExecute() function is used for SQL statements that do not return result sets, such as creating tables or views and inserting records.

Let's start by using DBI::dbGetQuery() to query the dogs_table_from_df table we created in our database earlier. For our query, let's find the top 10 zip codes ranked by the number of dog registrations in their dataset. To do this, we'll group the records in this table by the ZipCode column, and then use the DuckDB count aggregation function to get the number of registrations in each zip code:

```
dbGetQuery(
    conn,
    "
    SELECT ZipCode,
        count(*) AS num_registrations
    FROM dogs_table_from_df
    GROUP BY ZipCode
    ORDER BY num_registrations DESC
    LIMIT 10
    "
)
```

This produces the following dataframe:

ZipCode	num_registrations
<dbl>	<dbl>
10025	13819
10023	11189
11201	10907
11215	10849
10024	10581
10011	10249
10128	10174
10009	9678
10314	8886
10312	8805

A data.frame: 10 × 2

Figure 9.4 – A dataframe showing the top 10 zip codes ranked by the number of dog registrations

We can see that the DBI::dbGetQuery() function executes immediately, returning its results as a dataframe. This example illustrates a simple workflow for evaluating a custom SQL query against a DuckDB database and getting its results back as a dataframe, which we can continue to work with in R.

If you are working interactively with R, such as when performing data analysis, you will almost always want to use DBI::dbGetQuery() to evaluate SQL queries synchronously. However, there are times when you may want to retrieve query results asynchronously, such as when developing an application or library. For these situations, consult the DBI documentation for the DBI::dbSendQuery() and DBI::dbFetch() functions: https://dbi.r-dbi.org/reference/dbSendQuery.html.

When you need to run a SQL statement that does not return a result set, the function from the DBI package you want is DBI::dbExecute(). This takes the same arguments as DBI::dbGetQuery() – a DBI connection and string containing a SQL statement – with the difference being that it does not return a dataframe. Let's have a look at a situation where we would want to use DBI::dbExecute().

Let's look at how we might load the NYC Dog Licensing dataset from the original CSV file we started with using DuckDB.

Imagine that we have a workflow where new versions of the NYC Dog Licensing CSV file are regularly updated. We want to ensure that we can query the latest version of this file from DuckDB, without needing to continually update a table. To meet this need, we'll create a view in the database that can be reused for ingesting CSV snapshots of this dataset. Recall from *Chapter 3* that a SQL view allows us to create an entry in our database's catalog that is associated with a query. This allows us to query the view as if it were a table, with the underlying query of the view being executed each time the view is queried.

Since creating a view does not return a result set, we'll use `DBI::dbExecute()` to run a SQL statement that creates a view containing the DuckDB query that will load the records in our CSV file. Here's the code to do this:

```
dbExecute(
    conn,
    "
    CREATE OR REPLACE VIEW nyc_dogs_csv_view AS
    SELECT *
    FROM read_csv(
        'NYC_Dog_Licensing_Dataset.csv',
        ignore_errors=True
    )
    "
)
0
```

The number that's returned as output indicates the number of rows that were loaded into DuckDB. In this case, this is 0 since we have created a view.

Also, note that we used the `ignore_errors` parameter of DuckDB's `read_csv` function. As we saw in *Chapter 2*, this instructs DuckDB's CSV reader to skip over records in the CSV that cannot be parsed according to the schema being used for reading. In this case, this is the schema that's automatically inferred by the CSV sniffer as we haven't manually specified one using the `columns` parameter. We've used this because the snapshot of our dataset has some data quality issues, with some malformed records that need to be skipped; otherwise, queries from this view would fail with a CSV parsing error.

We now have a reusable view that, when queried, will retrieve the parsed contents of the CSV file. If the CSV file's contents change as new versions of the dataset become available, queries to our view will always retrieve the latest data. Let's check that this view works by sampling three records from the CSV file and retrieving them as a dataframe using `DBI::dbGetQuery()`:

```
dbGetQuery(
    conn,
    "
    SELECT *
```

```
    FROM nyc_dogs_csv_view
    USING SAMPLE 3
    "
)
```

This produces a dataframe that contains the three records that were randomly sampled from the CSV file:

AnimalName	AnimalGender	AnimalBirthYear	BreedName	ZipCode	LicenseIssuedDate	LicenseExpiredDate	Extract Year
<chr>	<chr>	<dbl>	<chr>	<dbl>	<date>	<date>	<dbl>
BIBI	F	2016	Chow Chow	10036	2022-06-09	2027-06-09	2023
PENNY	F	2020	Goldendoodle	10023	2020-04-01	2021-04-01	2022
COCO	F	2013	Terrier mix	10019	2017-12-22	2018-12-22	2017

A data.frame: 3 × 8

Figure 9.5 – A dataframe with three records sampled from nyc_dogs_csv_view

Since we are taking a random sample from this view, the three records you'll see in your output will be different from those displayed here. Something worth noting from this example is that DuckDB's CSV sniffer was able to detect and parse both date fields as DuckDB DATE types, despite their less standard format. This is in contrast to the readr::read_csv() R function, which needed to be given instructions on how to parse this date format, illustrating the impressive auto-detection capabilities of DuckDB's CSV reader. One difference in our resulting dataframe is that the AnimalGender column is not a facet containing the categorical gender values. In DuckDB, a categorical column like this would correspond to an ENUM data type, which would have the possible values of F and M. However, DuckDB's sniffer is not currently capable of inferring ENUM data types. While the use of an ENUM column would introduce some efficiency improvements, this isn't a problem for our purposes. Overall, the switch to using DuckDB's CSV reader, combined with the use of a reusable view, has simplified our ingestion code and workflow when compared to the readr::read_delim() approach we used previously. With this view created, we can use it through both SQL directly and through DBI methods that reference tables. For example, to retrieve the contents of the view as a dataframe to work with this dataset in R, we can either query the view by using a SQL query with DBI::dbGetQuery() or use the DBI::dbReadTable() function we saw previously to read the entire contents of the view. The following example shows these two distinct ways of retrieving a dataframe containing the complete results of nyc_dogs_csv_view:

```
dogs_df1 = dbGetQuery(
    conn,
    "SELECT * FROM nyc_dogs_csv_view"
)
dogs_df2 = dbReadTable(conn, "nyc_dogs_csv_view")
```

In terms of choosing between these two strategies, `DBI::dbReadTable()` offers a simple way of reading the complete contents of a database table or view into a dataframe, whereas `DBI:dbGetQuery()` allows you to provide a SQL query whose results will be loaded into a dataframe.

Using prepared statements

The DBI library provides support for prepared statements, which allows you to safely parameterize SQL queries and statements without exposing yourself to SQL injection vulnerabilities. Both `DBI::dbGetQuery()` and `DBI::dbExecute()` can take a second argument. This is a list of parameters that will be matched with the corresponding ? characters in the SQL query. For example, we can parameterize a query of the `nyc_dogs` table as follows:

```
dbGetQuery(
    conn,
    "
    SELECT *
    FROM nyc_dogs_csv_view
    WHERE BreedName = ? AND AnimalGender = ?
    LIMIT ?
    ",
    list("Beagle", "M", 3)
)
```

This produces the following dataframe:

AnimalName	AnimalGender	AnimalBirthYear	BreedName	ZipCode	LicenseIssuedDate	LicenseExpiredDate	Extract Year
<chr>	<chr>	<dbl>	<chr>	<dbl>	<date>	<date>	<dbl>
SAMMY	M	2007	Beagle	11231	2014-09-14	2017-10-15	2016
BAXTER	M	2013	Beagle	11374	2014-09-15	2019-11-21	2016
JUPLAY	M	2002	Beagle	11106	2014-09-16	2017-10-23	2016

A data.frame: 3 × 8

Figure 9.6 – A dataframe that contains the results of the prepared query

In this query, we passed in a list whose values are matched with the two slots in the `WHERE` clause, and the one slot in the `LIMIT` clause.

You can also create prepared statements with placeholder slots that can be reused multiple times as needed. To do this, you'll want to use `DBI::dbSendQuery()` for queries that return result sets and `DBI::dbSendStatement()` for other statements, such as deletions, insertions, and creations. In each case, you will create a result object, after which you can use `DBI::dbBind()` to invoke a specific set of values multiple times. Once you've used this result object, you need to use `DBI::dbClearResult()` to clean up the result object.

Let's go through an example of using `DBI::dbSendStatement()` to insert multiple values into the `dogs_table_from_df` table in our database:

```
result <- dbSendStatement(
    conn,
    "INSERT INTO dogs_table_from_df VALUES (
        ?, ?, ?, ?, ?, ?, ?, ?)"
)
dbBind(
    result,
    list(
        "John", "M", 2013, "Jack Russell Terrier",
        10261, "2014-07-12", "2017-08-09", 2016
    )
)
dbBind(
    result,
    list(
        "Lady Fluffina", "F", 2014, "Bichon Frisé",
        10302, "2014-08-22", "2017-09-25", 2016
    )
)
dbClearResult(result)
```

In this example, we created a result object that contains a prepared insertion statement for creating a new dog registration record, with slots for each field in the table. Then, we used `DBI::dbBind()` to insert two new records in this table by binding the slots in the insert statement with custom field values. Finally, we cleaned up the result object using the `DBI::dbClearResult()` function.

Let's confirm that we wrote our new records successfully:

```
tail(dbReadTable(conn, "dogs_table_from_df"), 2)
```

This produces our two records in a dataframe:

	AnimalName	AnimalGender	AnimalBirthYear	BreedName	ZipCode	LicenseIssuedDate	LicenseExpiredDate	Extract.Year
	<chr>	<fct>	<dbl>	<chr>	<dbl>	<date>	<date>	<dbl>
616891	John	M	2013	Jack Russell Terrier	10261	2014-07-12	2017-08-09	2016
616892	Lady Fluffina	F	2014	Bichon Frisé	10302	2014-08-22	2017-09-25	2016

A data.frame: 2 × 8

Figure 9.7 – A dataframe showing the two records that we inserted
into the database using a prepared query

Now that we've made some changes to our database, we'll finish up our tour of using the DBI package to work with DuckDB by seeing how we can close our database connections safely.

Disconnecting from DuckDB

When you've finished working with your DuckDB database, it's important to close the connection. This is especially true when you're writing to a persistent on-disk database, to ensure that all writes are flushed from any temporary data structures to disk. While connections are closed implicitly when they go out of scope, to avoid ambiguity and risk of data loss, it is safer to close connections explicitly using the `DBI:dbDisconnect()` function:

```
dbDisconnect(conn, shutdown = TRUE)
```

In this snippet, we closed the connection we've been working on that's associated with the `conn` variable. We also used the `shutdown` argument to ensure that the DuckDB database we had an open connection to was shut down at the same time. Closing the database when closing the connection object behavior is almost always what you want to do, and this will likely become the default behavior of DuckDB's `DBI:dbDisconnect()` implementation in the future.

If you find yourself defining a function that needs to establish a connection to a DuckDB database for the duration of the function, a good practice is to set up an **exit handler**, which is an R feature that allows you to register code that will be run, regardless of whether the function ends normally or as the result of an error. This is particularly useful when you're setting up any temporary global state, such as a database connection, that should be reset when it is no longer needed. You can do this using R's `on.exit()` function, which takes an expression as an argument that will be run when the function exits. Let's say we wanted to define a function called `run_query()` that takes a DuckDB database path and a SQL query string as arguments and evaluates the SQL query against the target database, returning the results as a dataframe. Inside this function, after opening a connection to the target database, we can add an exit handler that will close the connection and shut down the database on completion of the function. This is how this would look:

```
run_query <- function(db_path, sql_query) {
    conn <- dbConnect(duckdb::duckdb(), dbdir = db_path)
    on.exit(dbDisconnect(conn, shutdown = TRUE), add = TRUE)
    dbGetQuery(conn, sql_query)
}
```

Note that we also set the `add = TRUE` argument to `on.exit()`, which is almost always what you will want to do. This causes a new exit handler to be added, rather than overwriting a previous exit handler. Even if you only have one function with an exit handler now, getting in the habit of doing this will prevent any unwanted surprises later when you start adding additional exit handlers.

Since we have been working with an on-disk DuckDB database at the `quack.duckdb` file path, we can run our new function as follows:

```
run_query(
    "nyc_dogs.duckdb",
    "SELECT * FROM dogs_table_from_df USING SAMPLE 3"
)
```

This gives us the following dataframe:

AnimalName	AnimalGender	AnimalBirthYear	BreedName	ZipCode	LicenseIssuedDate	LicenseExpiredDate	Extract Year
<chr>	<fct>	<dbl>	<chr>	<dbl>	<date>	<date>	<dbl>
APPLE	F	2011	Chihuahua	11231	2018-12-02	2020-08-25	2018
BLUEBELL	F	2014	Siberian Husky	11225	2020-04-28	2021-03-18	2022
DIXIE	F	2015	Terrier mix	11101	2023-06-15	2024-06-18	2023

A data.frame: 3 × 8

Figure 9.8 – A dataframe showing the three records sampled from the nyc_
dogs.duckdb database after using our R run_query() function

By passing our target database and target query into the `run_query()` function, a connection to the database is created and then automatically closed when the function returns. If we need to add further behavior to this function, we can safely extend the code without having to remember to close the connection at the end.

In this section, while exploring how to work with DuckDB using R's DBI, we covered using DuckDB to query tables that we created out of a dataframe, and we also queried directly from a source CSV file. In the next section, we'll look at how we can query both dataframes and Arrow tables directly.

Registering R objects as virtual tables

An alternative approach to querying data from a dataframe, without needing to load it into a new table first, is to register a dataframe or tibble as a virtual table. This is similar to how a SQL view works, in that it involves creating an entry in the database catalog that points to a target dataframe, enabling DuckDB queries to be performed against the dataframe directly. We can also do the same thing with Apache Arrow tables, registering them as views within our DuckDB database's catalog. In this section, we'll provide some brief examples for registering both dataframes and Arrow tables as views that we can query.

We'll use the `nyc_dogs.duckdb` database we created previously for this section. Since we closed our previous connection to this database, we'll need to create a new one:

```
conn <- dbConnect(
    duckdb::duckdb(),
    dbdir = "nyc_dogs.duckdb"
)
```

Now, we're ready to register some R objects as virtual tables in DuckDB.

Registering a dataframe as a virtual table

To register a dataframe or tibble as a virtual table, we can use the `duckdb::duckdb_register()` function. Let's see how this works with the `dogs_df` tibble we created in the previous section:

```
duckdb::duckdb_register(conn, "dogs_df_view", dogs_df)
dbListTables(conn)
```

We get the following result:

```
'dogs_df_view'·'dogs_table_from_df'·'empty_dogs_table'·'nyc_dogs_csv_
view'
```

Our tibble is now registered as a virtual table called `dogs_df_view`, which we have confirmed by calling `DBI::dbListTables()` to see the contents of the database catalog. Note that, just as if we had created a DuckDB view using `CREATE VIEW`, no data has been read into our DuckDB database yet. We are now able to query this virtual table backed by our tibble as if it were a standard table:

```
dbGetQuery(
    conn,
    "SELECT * FROM dogs_df_view USING SAMPLE 3"
)
```

This gives us the following dataframe:

AnimalName	AnimalGender	AnimalBirthYear	BreedName	ZipCode	LicenseIssuedDate	LicenseExpiredDate	Extract Year
<chr>	<fct>	<dbl>	<chr>	<dbl>	<date>	<date>	<dbl>
MAGGIE	F	2008	Beagle	10128	2021-06-04	2022-06-04	2022
BAILEY	M	2013	Unknown	10471	2020-02-21	2021-02-27	2022
SCOUT	M	2020	Golden Retriever	11205	2022-03-01	2023-12-14	2023

A data.frame: 3 × 8

Figure 9.9 – A dataframe with three records sampled from an R
dataframe that's registered as a virtual table in DuckDB

You don't need to worry about the dataframe object remaining available to your program as DuckDB will automatically keep a reference to the dataframe to prevent it from being garbage-collected. DuckDB will then remove this reference when the connection you registered it on is closed. If you do need to manually clear the reference to allow the dataframe to be garbage collected and free up memory, you can explicitly unregister the virtual table using `duckdb::duckdb_unregister()`:

```
duckdb::duckdb_unregister(conn, "dogs_df_view")
dbListTables(conn)
```

We get the outputs as follows:

```
'dogs_table_from_df'·'empty_dogs_table'·'nyc_dogs_csv_view'
```

From the output of the `DBI::dbListTables()` call, we can see that the `dogs_df_view` entry has been removed from our DuckDB catalog.

Registering an Apache Arrow table as a virtual table

DuckDB also allows us to register Arrow tables as virtual tables. Arrow tables are in-memory column-oriented data structures that are similar to R dataframes, but since they are backed by the language-agnostic Apache Arrow format, they have the advantage of not needing to be serialized into memory across different language clients. If you want to learn more about leveraging Apache Arrow for fast and efficient data analytics, the book *In-Memory Analytics with Apache Arrow*, also published by Packt, is an excellent resource.

First, we'll need to load the `arrow` package. Note that when you load this package, you might get a warning about conflicting function names from a previously loaded package. We won't be using these, so you can ignore this message:

```
library(arrow)
```

To register an Arrow table as a virtual table, we need to use the `duckdb::duckdb_register_arrow()` function, which otherwise works the same as before for dataframes. We'll demonstrate this by using an Arrow table that's created by converting our `dogs_df` tibble into an Arrow table via the `arrow:arrow_table()` function:

```
dogs_arrow = arrow::arrow_table(dogs_df)
```

Now that we have an Arrow table, let's register it as a virtual table in our database:

```
duckdb::duckdb_register_arrow(
    conn,
    "dogs_arrow_view",
    dogs_arrow
)
```

With this Arrow table registered as a virtual table in our DuckDB database, we can query it just as usual:

```
dbGetQuery(
    conn,
    "SELECT * FROM dogs_arrow_view USING SAMPLE 3"
)
```

This produces the following dataframe:

AnimalName	AnimalGender	AnimalBirthYear	BreedName	ZipCode	LicenseIssuedDate	LicenseExpiredDate	Extract Year
<chr>	<chr>	<dbl>	<chr>	<dbl>	<date>	<date>	<dbl>
LUCY	F	2014	Siberian Husky	11234	2020-07-26	2021-06-06	2022
BROWNIE	F	2007	Poodle, Miniature	11420	2020-08-20	2021-03-29	2022
KINGSLY	M	2013	American Pit Bull Terrier/Pit Bull	10029	2020-04-30	2025-05-04	2023

A data.frame: 3 × 8

Figure 9.10 – A dataframe with three records sampled from an Arrow
table that's registered as a virtual table in DuckDB

Note that for Arrow-backed virtual tables, to discover those that have already been registered, you need to use the `duckdb::duckdb_list_arrow()` function:

```
duckdb::duckdb_list_arrow(conn)
```

We get the following result:

```
'dogs_arrow_view'
```

To unregister Arrow-backed virtual tables, you need to invoke the `duckdb::duckdb_unregister_arrow()` function:

```
duckdb::duckdb_unregister_arrow(conn, "dogs_arrow")
```

What about replacement scans?

In *Chapter 8*, we saw how the DuckDB Python client allows us to leverage replacement scans to query in-memory Python data structures, including dataframes and Arrow tables, as if they were tables in a DuckDB database, simply by referencing the name of variables in scope, and without having to explicitly register them. You might be wondering why we haven't mentioned this convenient feature in the context of the R client. At the time of writing this has not been implemented, however, work has commenced in that direction. Check the DuckDB documentation for updates!

Now that we've looked at a range of ways that we can register data sources with DuckDB for querying, we'll turn to look at how we can use the R's dplyr package to query DuckDB data sources more conveniently.

Querying DuckDB with dplyr

The dplyr package is highly regarded among data practitioners who use R for performing data analysis and modeling. It provides users with a set of key verbs for manipulating data, such as `select`, `filter`, `arrange`, `summarize`, and `mutate`. By enabling users to combine these verbs through a composable grammar of data manipulation, the dplyr API provides an elegant and intuitive interface for constructing analytical queries programmatically.

dplyr can be used to query a range of data backends, including R dataframes, Apache Arrow tables, Apache Spark datasets, and a variety of popular SQL databases. The dataframe backend is the most frequently used, allowing users to query R dataframes and tibbles using the dplyr interface. The dbplyr package provides an alternative backend that enables dplyr to be used as a query interface for a range of SQL-based databases. It works behind the scenes by translating dplyr operations into the SQL dialect of the database you have connected to. DuckDB has good support for dbplyr, which means that we can use dplyr to compose queries that are executed against DuckDB databases, without the need to write SQL.

> **duckplyr**
>
> While this book was in the final stages of completion, a new package called duckplyr was released. duckplyr is designed to be a drop-in replacement for dplyr when querying DuckDB databases. Rather than translating dplyr expressions into DuckDB SQL, duckplyr uses a DuckDB-native interface. This offers several benefits compared to using the standard dplyr package, including cleaner integration with DuckDB, improved query performance, and better diagnostic feedback around query errors. If you find the dplyr interface for working with DuckDB compelling, you may want to try out duckplyr. We'll also discuss duckplyr in *Chapter 12*, as we explore the wider DuckDB ecosystem.

In this section, we'll cover how we can take advantage of the convenient and elegant dplyr query interface to query DuckDB databases. We'll start by covering the foundations of how to use dplyr to query dataframes – an approach that works with any dataframe or tibble. We'll then explore how the dbplyr backend allows us to use the same dplyr interface to query DuckDB databases directly without needing to write any SQL. We'll do this by performing some analyses of the New York City dog registrations dataset. Specifically, we'll use the view we created in our `nyc_dogs.duckdb` database, which reads the parsed contents of the source CSV file.

Using dplyr to query dataframes

Before we use dplyr with DuckDB, let's start by seeing dplyr in action, using it to query a dataframe. We'll use the dogs_df dataframe we initially loaded from the NYC_Dog_Licensing_Dataset. csv file earlier in this chapter. Let's use dplyr to count the number of dog registrations for each year and rank them by year. Here's a dplyr query that combines the right verbs to achieve this:

```
dogs_df |>
    group_by(LicenseIssuedYear = year(LicenseIssuedDate)) |>
    summarise(count = n()) |>
    arrange(desc(count))
```

This gives us the following dataframe containing the number of dog registrations for each year in the dataset. This shows us that 2020 had the most registrations:

LicenseIssuedYear	count
<dbl>	<int>
2020	105090
2021	87246
2016	77015
2019	75087
2017	74282
2018	72513
2023	48106
2022	41282
2015	34712
2014	1557

A tibble: 10 × 2

Figure 9.11 – A tibble containing the number of dog registrations in NYC by year

To arrive at this summarized , we applied a sequence of dplyr operations to our initial dataframe. Each operation is separated by the |> operator, which is the R pipe operator. The pipe operator takes the output of the expression on its left-hand side and places it as the first argument of the function on the right-hand side, which, for all dplyr functions – and some base R functions, such as utils::head() – is the data argument. This allows us to elegantly compose a series of operations, with the input and outputs of each step flowing through, without having to explicitly pass through the data parameter.

Let's go through the sequence of operations in our dplyr query:

1. We used the `dplyr::group_by()` function to break the dataframe into groups of records with the same year of registration. We did this by defining a `LicenseIssuedYear` grouping key, whose value we derive for each group by `LicenseIssuedDate` by applying the `lubridate::year()` function (from the tidyverse's lubridate package) to the `LicenseIssuedDate` column.

2. After grouping a dataset, we typically need to apply a verb to each resulting group. In this case, we used the `dplyr::summarise()` function to collapse each of the groups into a single record containing the number of records in the group. To do this, we summarized each group via the `dplyr::n()` function, which returns each group's size, and then named this new summarized column `count`.

3. The last thing we needed to do was sort the years in descending order of registration counts, which we did using the `dplyr::arrange()` and `dplyr::desc()` functions.

From the perspective of the analysis we have performed, it's worth pointing out that these registration counts we're seeing don't represent the complete number of dog registrations for these years in NYC. This is because the dataset only contains active registrations for each year of extract. At the present time of analysis, this dataset contains extracts from the years 2016-2022, which is to say, it only contains records for active dog registrations during each of those years.

Back to our `dplyr` code, let's look at how it turns out that we can simplify the code for performing this analysis. Counting the occurrences of values in a table is such a common type of group-by and summarized set of operations that the `dplyr::count()` function exists to cater to this need. Let's use it to make our preceding query more concise:

```
dogs_df |>
    count(
        LicenseIssuedYear = year(LicenseIssuedDate),
        name = "count",
        sort = TRUE
    )
```

The resulting tibble from this query has the same contents as the one we saw in *Figure 9.11*. To match the same behavior as our previous solution, we used the `dplyr::count()` function's name parameter to specify the name of the new column with group sizes, as well as rank the output in descending order via the `sort` parameter. With these in place, our query is now much more concise thanks to the `dplyr::count()` function, further illustrating the expressive power of the dplyr interface.

> **Why are we using the |> pipe instead of the %>% pipe?**
>
> If you've used dplyr before, you may be wondering why we're using the | > pipe operator, which is the base R pipe operator, to compose our dplyr operations, rather than the `%>%` pipe operator, which comes from the `magrittr` package in the tidyverse. These two pipe operators largely have the same functionality, but the base R pipe operator is a more recent addition, only being introduced in R 4.1.0. The `%>%` operator does support some advanced functionality beyond the base R pipe operator; however, this isn't needed for the examples in this chapter. Hence, we've chosen to use the base R | > operator.

At this point, we haven't used dplyr to query DuckDB; we have just used it to query a dataframe that we extracted from our DuckDB database The approach of extracting entire tables from our database into dataframes to do analysis in R may work for small datasets, such as the one we're using now. However, as we discussed in *Chapter 1*, as the size of our dataset grows, we will start to run into issues, such as running out of memory or operations taking an unacceptably long amount of time to run. In the next section, we'll look at how we can leverage the dplyr interface to query the contents of DuckDB databases directly, allowing us to take advantage of DuckDB's superior efficiency and performance in our dplyr-based workflows.

Using dplyr to query DuckDB tables and views via dbplyr

dbplyr is an alternative backend for dplyr that translates dplyr code into a range of target SQL dialects. DuckDB has good support for dbplyr, which means that we can compose DuckDB queries using the convenient dplyr interface without the need to write SQL directly while still retaining all the performance benefits that DuckDB has over dataframes.

When using the dbplyr backend to query a SQL table, you will generally start by using the `dplyr::tbl()` function to create a lazy representation of a target table that we'll use to build up our query. This also works for views defined in your database's catalog. Let's do this for the view we created earlier:

```
tbl(conn, "nyc_dogs_csv_view")
```

This gives us a preview of the results our view returns:

```
# Source:    table<nyc_dogs_csv_view> [?? x 8]
# Database: DuckDB v0.10.2 [ned@Linux 6.7.9-060709-generic:R 4.3.2//home/ned/projects/duckdb/boo
k/Getting-Started-with-DuckDB/chapter_09/nyc_dogs.duckdb]
   AnimalName AnimalGender AnimalBirthYear BreedName    ZipCode LicenseIssuedDate
   <chr>      <chr>                  <dbl> <chr>          <dbl> <date>
 1 PAIGE      F                       2014 American P…    10035 2014-09-12
 2 YOGI       M                       2010 Boxer          10465 2014-09-12
 3 ALI        M                       2014 Basenji        10013 2014-09-12
 4 QUEEN      F                       2013 Akita Cros…    10013 2014-09-12
 5 LOLA       F                       2009 Maltese        10028 2014-09-12
 6 IAN        M                       2006 Unknown        10013 2014-09-12
 7 BUDDY      M                       2008 Unknown        10025 2014-09-12
 8 CHEWBACCA  F                       2012 Labrador R…    10013 2014-09-12
 9 HEIDI-BO   F                       2007 Dachshund …    11215 2014-09-13
10 MASSIMO    M                       2009 Bull Dog, …    11201 2014-09-13
# i more rows
# i 2 more variables: LicenseExpiredDate <date>, `Extract Year` <dbl>
```

Figure 9.12 – The display representation of a dbplyr lazy table object, showing a preview of its contents

Even though we can already see a few records from our view, only a preview of the table's contents has been read from the view so far. What we've created using the `dbplyr::tbl()` function is a lazily evaluated representation of our target table, which won't be fully evaluated until needed.

The central idea behind the dbplyr backend is to use this lazy table representation as our data source in our dplyr pipeline and to construct our desired query by applying a sequence of dplyr verbs. The result of this will still be a lazy table object, but it will now include our desired query logic. To convert this table object into its corresponding DuckDB SQL and run it against the target database, we need to call the `dplyr::collect()` function against the table object we have built up. We can do this by putting it at the end of our dplyr pipeline. Let's see this in action by applying a single-head operation to our view:

```
tbl(conn, "nyc_dogs_csv_view") |>
    head(3) |>
    collect()
```

We successfully retrieved a tibble that contains the results of the simple DuckDB query that we created using the dplyr interface, augmented with the dbplyr backend. To inspect the DuckDB SQL code that's generated by our query, we can replace `dplyr::collect()` at the end of the chain with `dplyr::show_query()`:

```
tbl(conn, "nyc_dogs_csv_view") |>
    head(3) |>
    show_query()
```

The output is as follows:

```
<SQL>
SELECT nyc_dogs_csv_view.*
FROM nyc_dogs_csv_view
LIMIT 3
```

Here, we can see the resultant SQL for this simple query, which simply selects all columns in the view, limiting the results to three rows.

Data wrangling with dplyr

Now that we've covered the dbplyr fundamentals, let's go deeper by jumping into some more analysis with the NYC dog registration dataset. In particular, we'll see if we can observe any interesting insights around dog names and breeds for dogs registered in NYC.

The first thing we'll need to do is transform our dataset so that we only have a single record for each dog. Without this step, we'll have multiple records for each dog due to repeat registrations. While no column forms a unique key, we can use a composite key made of the following fields to remove duplicate records: AnimalName, AnimalGender, BreedName, and ZipCode. We may have a few unfortunate casualties in cases where there are multiple dogs in the one zip code with the same name, breed, and sex, who will be collapsed into one row, but this is probably good enough for our purposes.

To reduce our dataset down, we'll start by using dplyr::filter() to remove dog names with values representing unknown dog names since these will not be resolved into distinct records in our new dataset. We'll do the same again for missing BreedName values. We'll then apply the dplyr::distinct() function to the columns we want to use as the composite key for our new dataset, which will remove duplicate rows that share these column values, as well as drop columns that aren't included.

Here's our dplyr query:

```
unique_dogs <- tbl(conn, "nyc_dogs_csv_view") |>
    filter(
        !AnimalName %in% c(
            "UNKNOWN", "NAME NOT PROVIDED", "NAME", "NONE"
        )
    ) |>
    filter(
        !BreedName %in% c("Unknown", "Not Provided")
    ) |>
    distinct(
        AnimalName, AnimalGender, AnimalBirthYear,
        BreedName, ZipCode
    )
```

Now that we have a query object, `unique_dogs`, that contains a lazily evaluated query that removes duplicate dog registrations, we can work with this query object as a logical data source, reusing it across multiple fragments of analysis. Let's start by just getting the first five entries:

```
unique_dogs |>
    head(5)  |>
    collect()
```

This produces the following tibble:

AnimalName	AnimalGender	AnimalBirthYear	BreedName	ZipCode
<chr>	<chr>	<dbl>	<chr>	<dbl>
MAIS	M	2018	Beagle Crossbreed	10029
ZIZANIA	F	2018	Boxer	10024
BOOMER	M	2011	Labrador Retriever	10128
LENNY	M	2016	Chihuahua Crossbreed	10463
BELLA	F	2016	American Pit Bull Terrier/Pit Bull	10301

A tibble: 5 × 5

Figure 9.13 – A dataframe showing the first five records of our new dataset,
which contains one record for each dog in the dataset

Note that you should see a different set of rows in this dataframe than in our results here. This is because we didn't specify an order for our results in our dplyr query. If you're used to working dplyr to query dataframes, this might be surprising. While dataframes have an inherent ordering, the results of SQL queries are not guaranteed to be ordered unless you specify one with an ORDER BY clause. To achieve a deterministic ordering with our dplyr query, we would need to include a `dplyr::arrange()` call at the end of our dplyr pipeline that specifies all the columns we included in the `dplyr::distinct()` function. This won't be an issue for our analysis, so we haven't included it. You might want to try to apply this change yourself. You may also encounter some further data quality issues in the `AnimalName` column that we haven't filtered out and that you could further uplift our dplyr query with.

Now that we've established the data preparation part of our query pipeline, we can move on to using it to answer some analytical questions. Let's say we wanted to find the top 10 most frequent names for registered male dogs. We can achieve this by extending the dplyr chain with the `dplyr::filter()` and `dplyr::count()` functions. We'll save this extended query to a variable called `table_query` rather than immediately collecting and executing it against the database:

```
table_query <- unique_dogs |>
    filter(AnimalGender == "M") |>
    count(AnimalName, name = "num_dogs", sort = TRUE) |>
    head(10)
```

To construct the query that will answer our question about male dog names, we used dplyr to continue applying operations to our existing `unique_dogs` lazy table object, which contains our data preparation logic. First, we used the `dplyr::filter()` function to include only male dogs, after which we used the `dplyr::count()` function to count up the number of occurrences of each distinct name and rank the results in descending order.

Before we run this query, let's see what the SQL it generates looks like:

```
table_query |>
    show_query()
```

We can see the output as follows:

```
<SQL>
SELECT AnimalName, COUNT(*) AS num_dogs
FROM (
  SELECT DISTINCT AnimalName, AnimalGender,
    AnimalBirthYear, BreedName, ZipCode
  FROM nyc_dogs_csv_view
  WHERE
    (NOT(AnimalName IN ('UNKNOWN', 'NAME NOT PROVIDED',
      'NAME', 'NONE'))) AND
    (NOT(BreedName IN ('Unknown', 'Not Provided'))) AND
    (AnimalGender = 'M')
) q01
GROUP BY AnimalName
ORDER BY num_dogs DESC
LIMIT 10
```

Since our query has become more involved, so too has the SQL that it generates. Take a moment to compare this SQL with the dplyr query chain. See if you can identify which parts of the SQL query correspond to the different dplyr operations we used to construct our query. Remember that this query started by using the dogs query object as a logical data source. This means that the underlying query in `table_query` contains both the initial de-duplication logic and the filtering and counting operations. Here's an example:

- The `SELECT DISTINCT` statement comes from `dplyr::distinct()`
- The `WHERE` clause's expression comes from `dplyr::filter()`
- The aggregation components, `COUNT(*)` and `GROUP BY`, come from `dplyr::count()`
- The `ORDER BY` clause comes from the `sort = TRUE` parameter to `dplyr::count()`
- The `LIMIT` clause comes from the `utils::head()` function

With that covered, let's run our query against the database:

```
table_query |>
    collect()
```

This gives us the following tibble:

AnimalName	num_dogs
<chr>	<dbl>
MAX	2341
CHARLIE	1814
ROCKY	1600
MILO	1512
TEDDY	1452
BUDDY	1228
TOBY	1053
LEO	1047
LUCKY	1016
OLIVER	976

A tibble: 10 × 2

Figure 9.14 – A dataframe that contains the top 10 most frequent
names for registered male dogs in the dataset

We can finally see the top 10 most frequently occurring male dog names present in this dataset of NYC dog registrations, with *Max* being the most frequent.

Another line of investigation we could explore is around changes in popular dog names over time. A rather simple analysis could be to identify the most popular dog name each year. To do this, we need to group the dataset into birth years, and then for each year, count the occurrence of names, selecting the most frequently occurring name and the number of times it occurs. It turns out that while it is simple to describe with words, queries that involve selecting the row with the greatest value for a specific column from each group can be trickier to specify in SQL than you might expect, with solutions typically involving window functions or self-joins. This is where dplyr can help as its interface is frequently able to render complex queries more intuitively compared to their SQL equivalents.

For this analysis, let's restrict our analysis to names of female dogs. Additionally, since this dataset is constructed from active dog registrations from 2016 to 2022, we won't be able to make interesting observations about the earliest years in the dataset – the earlier back we go, the older and fewer dogs we'll see, giving us less confidence to make any observations. To handle this, we'll limit records to only dogs born after 2010.

Let's assemble the dplyr query we need to identify the most popular name for female dogs for each year, assigning the resulting query object to a variable named `pop_dog_names`:

```
pop_dog_names <- unique_dogs |>
    filter(AnimalGender == "F", AnimalBirthYear > 2010) |>
    count(AnimalBirthYear, AnimalName, name = "NumDogs") |>
    slice_max(by = AnimalBirthYear, order_by = NumDogs) |>
    arrange(AnimalBirthYear)
```

Let's go through each function in our dplyr pipeline:

1. We started by using `dplyr::filter()` to restrict records to female dogs and also those born after 2010.

2. Then, we used `dplyr::count()` to create a table with the `AnimalBirthYear`, `AnimalName`, and NumDogs columns, which enumerates pairs of year and dog names, along with the number of dogs sharing these values.

3. After, we needed to group that list by `AnimalBirthYear`, and then for each year extract the single record with the highest NumDogs value, which will correspond to the dog name that occurs most frequently for that year. Fortunately for us, `dplyr::slice_max()` can do all this in one step with the right configuration. Its by parameter allows us to perform the slicing across groups of rows with the same `AnimalBirthYear`; then, with its `order_by` parameter, we specify that the slicing within those groups is performed by taking the record with the maximum NumDogs value.

4. The final step we performed was to use `dplyr::arrange()` to sort our table of results by `AnimalBirthYear`.

Before we collect and run this somewhat complex query against the database to see the results, you might find it beneficial to inspect intermediate results at different stages in the dplyr pipeline. You can do this by calling dplyr::collect() on progressively larger sequences of dplyr operations. Once you feel comfortable with the logic of our query, we'll collect and evaluate our complete query:

```
pop_dog_names |>
    collect()
```

This produces the following tibble:

AnimalBirthYear	AnimalName	NumDogs
<dbl>	<chr>	<dbl>
2011	BELLA	146
2012	BELLA	138
2013	BELLA	169
2014	BELLA	191
2015	BELLA	198
2016	BELLA	236
2017	LUNA	236
2018	LUNA	241
2019	LUNA	251
2020	LUNA	298
2021	LUNA	261
2022	LUNA	151
2023	LUNA	50

A tibble: 13 × 3

Figure 9.15 – A dataframe containing the most frequently occurring names for female dogs by year

As desired, we now have a table that lists each year along with the name that occurred the most frequently, along with the number of dogs with that name.

This is quite a complex query that dplyr has enabled us to represent concisely. This would be somewhat verbose in SQL, making this a good example of how dplyr can improve the effectiveness of both creating and understanding queries. You might like to compare the dplyr code with its corresponding SQL by piping the pop_dog_names query object to dplyr::show_query() instead of dplyr::collect(). You can also go further than this by invoking the dplyr::explain() function instead. This will print the converted SQL, as well as print the result of submitting an EXPLAIN statement to the DuckDB database for that SQL query.

Calling DuckDB functions with dplyr

One of the convenient features of dplyr's interface is that we can readily combine R functions with its data manipulation verbs. We saw this earlier when we needed to count the number of dog registrations by year, but the column we had available was a full date. The count supports new derived columns to be created for the grouping by passing in column names and expressions to generate their contents. We created a new `LicenseIssuedYear` column for grouping by using the `lubridate::year()` function. We did this previously using the dataframe backend. Let's repeat this, but this time using dbplyr to query our DuckDB database:

```
issued_by_year = tbl(conn, "nyc_dogs_csv_view") |>
    count(
        LicenseIssuedYear = year(LicenseIssuedDate),
        name = "Count",
        sort = TRUE
    )
issued_by_year |>
    collect()
```

This produces the following tibble:

LicenseIssuedYear	Count
\<dbl\>	\<dbl\>
2020	105090
2021	87246
2016	77015
2019	75087
2017	74282
2018	72513
2023	48106
2022	41282
2015	34712
2014	1557

A tibble: 10 × 2

Figure 9.16 – A dataframe showing the number of license registrations in the dataset by year

Success! We were able to use dplyr to construct a query that produced the appropriate SQL to perform this transformation as part of the aggregation operation. Let's have a look at the SQL that was produced:

```
issued_by_year |>
    show_query()
```

We get the following output:

```
<SQL>
SELECT LicenseIssuedYear, COUNT(*) AS Count
FROM (
  SELECT
    nyc_dogs_csv_view.*,
    EXTRACT(year FROM LicenseIssuedDate) AS LicenseIssuedYear
  FROM nyc_dogs_csv_view
) q01
GROUP BY LicenseIssuedYear
ORDER BY Count DESC
```

As we can see, dbplyr knows that the R lubridate::year(LicenseIssuedDate) expression should map to the SQL extract(year FROM LicenseIssuedDate) expression. dbplyr is configured to translate a range of R functions into corresponding SQL expressions and operations; this includes support for lubridate functions. This SQL translation logic is made up of both standard SQL rules shared across all supported databases, as well as rules associated with supporting database-specific SQL dialects.

One of DuckDB's attractive features is its API's inclusion of a rich collection of functions that support a range of different analytical use cases. To make effective use of DuckDB with dplyr and dbplyr, we're going to want to be able to leverage these functions in our dplyr operation chains. Fortunately, dbplyr has us covered as it passes unknown function identifiers into the resultant generated SQL. This means that in many contexts, we can simply use DuckDB functions within our dplyr code as if they were R functions.

Let's look at an example of this in action. When working with variable-length strings, sometimes, we need to convert them into a fixed-size representation, typically an integer. This process is known as **hashing**. A commonly used hash function for converting strings into fixed-size integers is the MD5 algorithm. Let's imagine that we needed to generate a new column in our unique-dogs dataset that contains the MD5 hash of the concatenated AnimalName and ZipCode values for each record, which gives us a unique value for each pair of dog name and zip code values. DuckDB provides two functions that we can leverage for this task: md5 and concat. This means that our dplyr code can now be expressed through a simple call to the dplyr::mutate() function, which is used to create, modify, and remove columns in the output's results. Here, we want to add a new column called Hash, which we can do by providing it as a parameter to `dplyr::mutate()` and setting its value to the hash-producing expression:

```
unique_dogs |>
    mutate(Hash = md5(concat(AnimalName, ZipCode))) |>
    head(3) |>
    collect()
```

This gives us the following tibble:

AnimalName	AnimalGender	AnimalBirthYear	BreedName	ZipCode	Hash
<chr>	<chr>	<dbl>	<chr>	<dbl>	<chr>
GEORGE	M	2014	Italian Greyhound	10012	051abbf25f1611a65b239ee326338f69
HADDIX	F	2020	Pomeranian	11224	648784ffacfb8adec3c834fc76d36943
PETERS	F	2020	Yorkshire Terrier	11236	f8c4a5318b3a1a4cc9bd6bad8eb29a56

A tibble: 3 × 6

Figure 9.17 – A dataframe showing three dog records with a Hash column
derived from the MD5 hash of the combined dog name and zip code

With the help of the `dplyr::mutate()` function and DuckDB's `hash` and `concat` functions, capturing this transformation in dplyr can be performed rather concisely and elegantly.

Let's look at one more example of leveraging a DuckDB function in a dplyr query. This time, we'll come up with a query that allows us to look for all dog names that are similar to a target name. As part of its extensive collection of functions, DuckDB supports a range of text-similarity functions. You can find these here: `https://duckdb.org/docs/sql/functions/char.html#text-similarity-functions`. We'll use the **Jaro-Winkler distance**, which provides a measure of the distance between two strings in terms of how many edits it would take to transform one into the other. This distance takes real number values from 0 to 1, with 0 indicating an exact match and 1 indicating no similarity. DuckDB makes this available via the `jaro_winkler_similarity` function, which takes two strings arguments, returning their Jaro-Winkler distance. Another DuckDB function that will come in handy is `round`, which rounds numerical values to a desired number of decimal places.

For our query, we'll use the following functions to apply these steps in sequence:

- `dplyr::count()`: Counts the number of occurrences of each name
- `dplyr::filter()`: Drops names that occur less than 100 times
- `dplyr::mutate()`: Calculates and adds the Jaro-Winkler distance for each name against our target name, rounding the score to three decimal places
- `dplyr::arrange()`: Ranks the resultant words by highest edit-distance similarity
- `utils::head()`: Takes the top 10 words

Here's our dplyr chain that performs this query, comparing candidate names against the target string of 'BELLA', again building on our existing de-duplicating logic contained in the unique_dogs query object:

```
unique_dogs |>
    count(AnimalName, name = "Count") |>
    filter(Count >= 100) |>
    mutate(
        EditDistance = round(
            jaro_winkler_similarity(AnimalName, "BELLA"), 3
        )
    ) |>
    arrange(desc(EditDistance)) |>
    head(10) |>
    collect()
```

This produces the following tibble:

AnimalName	Count	EditDistance
<chr>	<dbl>	<dbl>
BELLA	2741	1.000
ELLA	324	0.933
BELLE	259	0.920
STELLA	820	0.822
DELILAH	135	0.790
BILLY	135	0.760
ISABELLA	114	0.742
ZELDA	136	0.733
ELLIE	413	0.733
BILLIE	144	0.730

A tibble: 10 × 3

Figure 9.18 – A dataframe showing the top 10 dog names by edit-distance similarity to the "BELLA" string

The combined query gives us a table containing the top 10 dog names that are most similar to the `"BELLA"` string by edit distance, with each row in the table containing the name, the number of times it occurs, and the Jaro-Winkler distance score for that word from the target word, rounded to three decimal places. You might find it interesting to run this query against other target names and inspect the resulting top-ranked words and their similarity scores. For more information about the Jaro-Winkler distance, its Wikipedia article offers a good overview: `https://en.wikipedia.org/wiki/Jaro%E2%80%93Winkler_distance`.

These examples have served to further illustrate the synergistic combination of the dplyr interface and the DuckDB SQL API. They also hopefully illustrate the value of being familiar with the types of functions that can be found in the DuckDB API, so that you're ready to supercharge your analytics workflows with the right functions at the right time. You can browse the complete DuckDB function reference here: `https://duckdb.org/docs/sql/functions/overview`.

Summary

In this chapter, we went through the DuckDB R API, covering all the aspects that are required so that you can start using DuckDB effectively in your R data analysis workflow.

After making sure that we had an environment up and running to work with R and DuckDB, we went through how to interact with DuckDB via the R DBI package, which included connecting to DuckDB, reading and writing tables to and from dataframes, querying DuckDB tables and executing SQL statements, using prepared statements, and disconnecting from DuckDB. We then looked at registering dataframes and Arrow tables as virtual tables in DuckDB's catalog, allowing us to query these R data structures directly from DuckDB. We finished off by looking at how we can leverage the dplyr interface for effective data manipulation, using it to query dataframes produced by DuckDB as well as querying DuckDB directly using the dbplyr backend, which converts dplyr code into SQL queries.

With that, you know about the different ways in which you can leverage DuckDB in your R-based analytical workflows, and you're also equipped with some guiding principles for how to make effective use of them.

In the next chapter, we'll switch gears a little by surveying some of DuckDB's SQL enhancements and features that are designed to improve the experience of writing analytical queries, and which will allow you to write more effective SQL. This will include SQL syntax enhancements that offer quality-of-life improvements, as well as DuckDB's support for language features that extend the functionality and flexibility of SQL queries.

10
Using DuckDB Effectively

So far, we've seen how DuckDB can load, transform, and analyze data from a variety of sources. DuckDB is not only a flexible and extendable database, but it has also been developed with great regard for usability.

The authors and maintainers of the DuckDB project have leveraged significant learnings from mature data management systems and have addressed some of the "paper cuts" and annoyances you may have experienced while using other database systems.

In this chapter, you will learn about time-saving shortcuts that you can use when querying and adding data to a DuckDB database. You will also gain experience with some novel approaches to joining tables – with exercises that involve using positional and temporal joins.

In this chapter, we're going to cover the following main topics:

- Selecting columns effectively
- Applying function chaining
- Using `INSERT` effectively
- Leveraging positional joins and temporal joins
- Recursive queries and macros

We will end this chapter by covering some tips and tricks for using DuckDB effectively.

Technical requirements

The exercises in this chapter can be run via the DuckDB **command-line interface** (**CLI**). The files and example data for these exercises are available at https://github.com/PacktPublishing/Getting-Started-with-DuckDB/tree/main/chapter_10.

Selecting columns effectively

If you are familiar with other database management systems, you may have developed some "SQL muscle memory" regarding making do with some of the more tedious parts of writing SQL. In this section, we are going to explore some of the DuckDB niceties – little shortcuts and tweaks to make your use of DuckDB easier.

For the first set of exercises, we are going to take to the mountains with some ski-related data exercises. We will explore a variety of ways to nominate the columns for our queries, and learn how to replace values and mechanisms to exclude unwanted columns.

Let's head to the slopes and start with some data about skiers:

```
CREATE OR REPLACE TABLE skiers AS
SELECT *
FROM read_csv('skiers.csv');
```

With our table created, we can query our list of skiers:

```
SELECT *
FROM skiers;
```

The preceding code will show all the rows and columns of the skiers table:

skier_first_name varchar	skier_last_name varchar	skier_age int64	skier_height int64	skier_helmet_color varchar	skier_bib_color varchar
Alice	Smith	12	152	red	black
Bob	Blaese	16	178	blue	yellow
Carol	Wilson	32	159	yellow	pink
Dan	Jones	52	182	red	yellow
Erin	Taylor	22	168	black	green
Frank	Williams	18	187	yellow	red
Grace	Miller	24	172	pink	black
Heidi	Johnson	22	178	yellow	yellow
Ivan	Brown	21	185	green	pink
Judy	Moore	27	160	red	black

Figure 10.1 – Selecting all the fields from the skiers table

A very common SQL shortcut is the SELECT * syntax. Like in many databases, * is a SQL shortcut that's interpreted by DuckDB as a request to return all the columns.

One small but very useful affordance of DuckDB is the use of a final trailing comma in a multi-column selection; this does not raise a syntax error. So, we can list all of our columns with trailing commas, including the final `skier_bib_color`, column:

```
SELECT skier_first_name,
    skier_last_name,
    skier_age,
    skier_height,
    skier_helmet_color,
    skier_bib_color,
FROM skiers;
```

This SQL is a valid DuckDB query that results in the following output:

skier_first_name varchar	skier_last_name varchar	skier_age int64	skier_height int64	skier_helmet_color varchar	skier_bib_color varchar
Alice	Smith	12	152	red	black
Bob	Blaese	16	178	blue	yellow
Carol	Wilson	32	159	yellow	pink
Dan	Jones	52	182	red	yellow
Erin	Taylor	22	168	black	green
Frank	Williams	18	187	yellow	red
Grace	Miller	24	172	pink	black
Heidi	Johnson	22	178	yellow	yellow
Ivan	Brown	21	185	green	pink
Judy	Moore	27	160	red	black

Figure 10.2 – Selecting all but the age from the skiers table

Although seemingly minor, DuckDB's tolerance for accepting trailing commas is a much-appreciated enhancement to conventional SQL syntax. We can rapidly copy and paste lines of SQL – and won't need to remove the final comma as we develop new queries. DuckDB has emphasized ease of use – and you certainly appreciate these niceties when you need to use other database systems.

Now, consider a situation where we wish to display all the columns apart from the age of the skier. A conventional SQL SELECT query displaying all the columns apart from the skiers' ages requires us to methodically list each column apart from the `age` column. We also need to ensure each column name is followed by a trailing comma – apart from the final column name.

DuckDB can also explicitly omit columns with the EXCLUDE syntax. To ignore the skiers' ages and heights, we can simply specify these as columns to exclude:

```
SELECT s.* EXCLUDE(skier_age, skier_height)
FROM skiers AS s;
```

Here's the output:

skier_first_name varchar	skier_last_name varchar	skier_helmet_color varchar	skier_bib_color varchar
Alice	Smith	red	black
Bob	Blaese	blue	yellow
Carol	Wilson	yellow	pink
Dan	Jones	red	yellow
Erin	Taylor	black	green
Frank	Williams	yellow	red
Grace	Miller	pink	black
Heidi	Johnson	yellow	yellow
Ivan	Brown	green	pink
Judy	Moore	red	black

Figure 10.3 – Excluding age and height from the skiers table

This is an efficient way to select all but a handful of columns. In our situation, we can display all the attributes of the skiers and simply exclude the fields that might be considered sensitive or personally identifiable.

DuckDB also has a syntax to apply changes columns in place with the REPLACE syntax. This is also useful when it's applied to a dynamic set of columns. For example, let's say we want to select all the columns of the skiers table, yet we wish to round the skiers' ages to the nearest 10 years and convert them into integers – that is, we wish to change both the value and data type:

```
SELECT s.*
    REPLACE(round(skier_age / 10) * 10)::INTEGER AS skier_age
FROM skiers AS s;
```

The preceding code will select all the columns, dynamically replacing the value of the skiers' ages:

skier_first_name varchar	skier_last_name varchar	skier_age int32	skier_height int64	skier_helmet_color varchar	skier_bib_color varchar
Alice	Smith	10	152	red	black
Bob	Blaese	20	178	blue	yellow
Carol	Wilson	30	159	yellow	pink
Dan	Jones	50	182	red	yellow
Erin	Taylor	20	168	black	green
Frank	Williams	20	187	yellow	red
Grace	Miller	20	172	pink	black
Heidi	Johnson	20	178	yellow	yellow
Ivan	Brown	20	185	green	pink
Judy	Moore	30	160	red	black

Figure 10.4 – Dynamically replacing the skiers' ages

Here, we have replaced `skiers_age` with a calculated value rounded to the nearest decade.

Let's introduce one additional DuckDB nicety – the COLUMNS selector. This allows us to use regular expressions when choosing column names. We can use the COLUMNS expression to return all column names that have a color attribute:

```
SELECT skier_first_name, COLUMNS('.*color$')
FROM skiers;
```

This returns `skier_first_name`, along with the fields matching the regular expression:

skier_first_name varchar	skier_helmet_color varchar	skier_bib_color varchar
Alice	red	black
Bob	blue	yellow
Carol	yellow	pink
Dan	red	yellow
Erin	black	green
Frank	yellow	red
Grace	pink	black
Heidi	yellow	yellow
Ivan	green	pink
Judy	red	black

Figure 10.5 – Columns selected with a regular expression

The regular expression within COLUMNS(`'.*color$'`) has matched the `skier_helmet_color` and `skier_bib_color` columns and returned the appropriate columns. The use of the COLUMNS filter is a powerful mechanism to dynamically select columns from a table or a view.

An interesting feature of the COLUMNS function is the ability to dynamically select columns to build a filter. We can create a WHERE clause that applies across multiple columns that match the search expression. For example, we can use the COLUMNS function to dynamically select the `color` columns and look for only the rows with a value of `yellow`. This will construct a WHERE predicate where all `color` columns must match the filter criteria:

```
SELECT skier_first_name, COLUMNS('.*color.*')
FROM skiers
WHERE COLUMNS('.*color.*') = 'yellow';
```

The result of this query shows the matching skier, dressed completely in yellow:

skier_first_name varchar	skier_helmet_color varchar	skier_bib_color varchar
Heidi	yellow	yellow

Figure 10.6 – Columns filtered with a regular expression

This is equivalent to combining the candidate columns in a where clause, such as `skier_helmet_color='yellow' AND skier_bib_color='yellow'`. This is a powerful capability where you can apply the same predicate to a large collection of fields, without the need to list each one individually.

In this section, we learned how to work with multiple columns in DuckDB efficiently. You could produce the same result with vanilla SQL, but you need to write a lot more code. The use of `REPLACE`, `EXCLUDE`, and `COLUMNS` gives us an expressive syntax for working with multiple columns effectively. Now, let's move on to some other DuckDB niceties by introducing function chaining.

Applying function chaining

In this section, we'll introduce **function chaining**, an elegant programming style for calling multiple scalar functions sequentially. Function chaining (sometimes termed **method chaining**) is a familiar technique in object-oriented programming languages – and DuckDB supports this elegant style to increase the readability of code. This will be familiar to users of Python. So, let's learn how to use function chaining in DuckDB SQL.

We will be using our `skiers` table from the previous section. Let's familiarize ourselves with the data available by querying the two name columns:

```
SELECT skier_first_name, skier_last_name
FROM skiers;
```

This returns the string fields shown in the following screenshot:

skier_first_name varchar	skier_last_name varchar
Alice	Smith
Bob	Blaese
Carol	Wilson
Dan	Jones
Erin	Taylor
Frank	Williams
Grace	Miller
Heidi	Johnson
Ivan	Brown
Judy	Moore

Figure 10.7 – The name fields of the skiers table

Let's say we want to format the names of our skiers by performing the following steps:

1. Concatenate `skier_first_name` and `skier_last_name` with a single space between the names by using the CONCAT_WS function.

2. Convert the name into uppercase with the UPPER function.

3. Left-pad the name string so that the entire name fills 20 characters, with the left gap filled with the > chevron, by using LPAD.

4. Append a full stop (.) at the end of each name with the CONCAT function.

We can write the following conventional SQL to perform these steps:

```sql
SELECT CONCAT(
  LPAD(
    UPPER(
      CONCAT_WS(' ', skier_first_name, skier_last_name))
      , 20
    , '>')
  , '.') AS skier_name
FROM skiers;
```

When run on our `skiers` table, this results in the following list of formatted skier names:

skier_name varchar
>>>>>>>>>ALICE SMITH.
>>>>>>>>>>BOB BLAESE.
>>>>>>>>CAROL WILSON.
>>>>>>>>>>DAN JONES.
>>>>>>>>>ERIN TAYLOR.
>>>>>>FRANK WILLIAMS.
>>>>>>>>GRACE MILLER.
>>>>>>>HEIDI JOHNSON.
>>>>>>>>>>IVAN BROWN.
>>>>>>>>>>JUDY MOORE.

Figure 10.8 – Formatted skier names using conventional SQL

This SQL produces the desired output; however, it is very hard to read. Note that we need to construct the query inside-out, with CONCAT_WS nested in the middle, stepping outward to the following function steps. Keeping track of the parameters for each function becomes complicated. The CONCAT function on the first line of the query has its (`'.'`) parameter on line 7 – quite inelegant!

Each of the functions we're using here is a scalar function – meaning each returns a single value per invocation. Let's see how function chaining can make this same task a little more elegant. In DuckDB, we can pass the result from the first CONCAT_WS into the UPPER function by simply using a dot (.) between the function calls and continue doing so for the LPAD and CONCAT functions.

An equivalent approach to formatting our skier names using function chaining is shown here:

```
SELECT CONCAT_WS(' ', skier_first_name, skier_last_name)
    .UPPER()
    .LPAD(20, '>')
    .CONCAT('.') AS skier_name
FROM skiers;
```

Each scalar function is invoked sequentially, yielding the same result as before:

```
         skier_name
         varchar

>>>>>>>>>ALICE SMITH.
>>>>>>>>>>BOB BLAESE.
>>>>>>>>CAROL WILSON.
>>>>>>>>>>>DAN JONES.
>>>>>>>>>ERIN TAYLOR.
>>>>>>FRANK WILLIAMS.
>>>>>>>>GRACE MILLER.
>>>>>>>HEIDI JOHNSON.
>>>>>>>>>>>IVAN BROWN.
>>>>>>>>>>JUDY MOORE.
```

Figure 10.9 – Formatted skier names using function chaining

The formatting of our skier names is the same as the pure SQL approach, but the code is much cleaner and easier to read. DuckDB supports this same simplification for every scalar function. We can use the dot (.) operator notation to chain any functions together.

With that, we have looked at some elegant shortcuts for creating DuckDB SQL queries. Now, let's look at some of the lesser-known SQL tips for inserting data.

Using INSERT effectively

For this section, we will create an index on our skiers table to ensure each skier's name is unique:

```
CREATE UNIQUE INDEX skier_unique
ON skiers (skier_first_name);
```

With this index created, we can be assured that skier_first_name is a unique value.

Imagine that we have a new skier called Kim who has a blue helmet. We can use the conventional INSERT syntax to list the columns and values:

```
INSERT INTO skiers(skier_first_name, skier_helmet_color)
SELECT 'Kim' AS skier_first_name, 'blue' AS skier_helmet_color;
```

This isn't very difficult, but we do need to ensure we keep the positional order of the column names (skier_first_name, skier_helmet_color) so that they match the order of the data provided ('Kim', 'blue'). This can become tedious if we are dealing with very wide tables with numerous columns and need to keep the ordering consistent.

DuckDB allows us to use the BY NAME directive to signify the names of the columns that should be used to match the destination table column names.

Let's see this in action by adding Liam to our skiers table. Note that we are not specifying the column names in the INSERT clause – instead, we are relying on the names of the columns to guide the location of each field:

```
INSERT INTO skiers BY NAME
SELECT 'green' AS skier_helmet_color,
    'red' AS skier_bib_color,
    'Liam' AS skier_first_name;
```

Let's check that Liam was added to the skiers table:

```
SELECT skier_first_name,
    skier_helmet_color,
    skier_bib_color
FROM skiers
WHERE skier_first_name = 'Liam';
```

This results in a single record:

skier_first_name varchar	skier_helmet_color varchar	skier_bib_color varchar
Liam	green	red

Figure 10.10 – A new skier, Liam, has been inserted

With that, we have added Liam to the skiers table, even though the ordering of the column names was different, and have matched the data to the correct fields.

Before we leave this section, let's introduce another elegant feature of DuckDB when inserting data – **automated field updating**. Recall that there is a unique index on the skiers table to ensure that skier_first_name is distinctive. Our skier, Liam, has just arrived wearing a black colored helmet. We want to ensure we have captured the helmet's color correctly in the skiers table.

We could go through the steps of checking if Liam is already in the table and then update the skier_helmet_color field. Instead, we'll use the INSERT OR REPLACE syntax:

```
INSERT OR REPLACE INTO skiers BY NAME
SELECT 'Liam' AS skier_first_name, 'black' AS skier_helmet_color;
```

This updates every column of the existing row of Liam in the `skiers` table to the new values specified in the to-be-inserted row. We can check the result by querying, like so:

```
SELECT skier_first_name,
    skier_helmet_color,
    skier_bib_color
FROM skiers
WHERE skier_first_name = 'Liam';
```

This results in the following output:

skier_first_name varchar	skier_helmet_color varchar	skier_bib_color varchar
Liam	black	red

Figure 10.11 – Skier Liam with an updated helmet color

It is worth noting that Liam's `skier_helmet_color` is `black`, and the `skier_bib_color` column (which was not in the `INSERT OR REPLACE` statement) remains in place with a value of `red`. `INSERT OR REPLACE` is a neat shortcut that provides a shorter SQL syntax than the alternative of writing a `MERGE` or `ON CONFLICT DO UPDATE` command.

Now that we've covered some helpful shortcuts for inserting data, let's look at an interesting feature of DuckDB: positional joins.

Leveraging positional joins

Relational tables in databases are traditionally considered unordered. SQL uses the `ORDER BY` clause to specify the ordering when data is retrieved. However, much of the data we process from other systems has an implicit order. There may be a natural ordering with a log file (with newer records appended to the end of the file) and files on disk, such as Parquet or CSV, can have a natural ordering of the file rows. It can be frustrating (and somewhat perplexing) to users of dataframes who are used to structures that preserve their rows to discover that relational tables are unordered.

DuckDB has the very useful `POSITIONAL` join, which acknowledges the implicit row numbers in each external table. This positional matching characteristic is frequently used by pandas dataframes when joining on the ordinal positioning of rows.

In this section, we will load implicitly ordered data from files on disk and introduce DuckDB's capability to join on the relative position of rows within a file.

Consider that skiers are leaving the top of the mountain in a competition. As the skiers leave the top of the mountain, their names and helmet colors are appended to the end of a text file. Their skiing techniques are scored by a judge watching them as they zip down the mountain. The judge doesn't know who the skiers are – they are simply judging the skiing technique of anonymous skiers. The order the skiers leave the top of the mountain (within the skiers.csv file) needs to be matched to the order the scores are entered by the judges (in the skier_scores.csv file). The ordering of rows within each file is important to ensure the correct skier is matched to the correct score.

We will start by loading two tables from disk. The skiers table contains a familiar list of skiers. We will recreate this by loading it from the skiers.csv file:

```
CREATE OR REPLACE TABLE skiers AS
SELECT *
FROM read_csv('skiers.csv');
```

Then, we will create the scores table by loading the skier_scores.csv file:

```
CREATE OR REPLACE TABLE scores AS
SELECT *
FROM read_csv('skier_scores.csv');
```

In many traditional relational database systems, the two tables would be considered unordered, and we would need to introduce another element to match the skiers to their scores. However, DuckDB can implicitly use the tables as two dataframes, connecting them using their ordering with POSITIONAL JOIN:

```
SELECT df1.skier_first_name,
    df1.skier_last_name,
    df2.score
FROM skiers AS df1 POSITIONAL JOIN scores AS df2;
```

This results in the following output:

skier_first_name varchar	skier_last_name varchar	score int64
Alice	Smith	8
Bob	Blaese	9
Carol	Wilson	4
Dan	Jones	7
Erin	Taylor	6
Frank	Williams	9
Grace	Miller	9
Heidi	Johnson	5
Ivan	Brown	7
Judy	Moore	8

Figure 10.12 – Positional join across two tables

The result of POSITIONAL JOIN gives us a query result that matches the skier to the correct score. It looks like we have a wide range of talent among our skiers. Now that we have seen how DuckDB can infer order from tables sourced from external files, we can look at some useful temporal functions available in DuckDB.

Using temporal joins with ASOF

Joining tables usually involves linking a common attribute across a table, such as finding the same identifier, name, or code value. DuckDB can also support "fuzzy" joins when we wish to join on values that may be close (but not identical) across tables. This is especially true when it comes to "temporal joins" – or joins that match on moments in time.

In this section, we will be discussing temporal joins and introducing the ASOF join, which can simplify joins across time.

For this exercise, we will be looking at historic weather measurements to see if bad weather is affecting our skier's performance. Let's start by loading data into the weather table:

```
CREATE OR REPLACE TABLE weather AS
SELECT *
FROM read_csv('weather.csv', timestampformat='%Y-%m-%d %H:%M:%S');
```

Let's take a peek at the data within our newly created `weather` table:

```
SELECT *
FROM weather
LIMIT 10;
```

Here's what the data looks like:

measurement_time timestamp	wind_speed int64	temp int64
2023-12-01 10:00:00	7	-9
2023-12-01 11:00:00	19	-8
2023-12-01 12:00:00	8	-6
2023-12-01 13:00:00	9	-3
2023-12-01 14:00:00	10	-2
2023-12-01 15:00:00	10	-2
2023-12-01 16:00:00	12	-2
2023-12-01 17:00:00	12	-3
2023-12-01 18:00:00	12	-3
2023-12-01 19:00:00	11	-4

Figure 10.13 – The weather table's sample rows

Here, we have a `measurement_time` field that specifies the timestamp where a windspeed and temperature measurement has been taken.

If we had a skier on the mountain who got judged for their ski run at `2023-12-01 10:01`, we can't know the precise weather then as we don't have a measurement at that exact moment in time.

As we only have hourly measurements for the weather, we might want to find the most recent measurement for wind and temperature at the time the skier was on the mountain. We can construct a query like this to find the most recent weather measurement for the time of the ski run:

```
WITH weather_cte AS(
    SELECT measurement_time,
    wind_speed,
    temp,
    LEAD(measurement_time, 1)
        OVER (ORDER BY measurement_time) AS measurement_end
    FROM weather
    ORDER BY measurement_time
)
```

```
SELECT *
FROM weather_cte
WHERE TIMESTAMP '2023-12-01 10:01:00'
    BETWEEN measurement_time AND measurement_end;
```

Quite a bit is going on in this query, so let's break it down. A **common table expression** (CTE) selects data from the `weather` table, calculating the next `measurement_time` using the LEAD function, ordered by `measurement_time`. We find the weather at the `2023-12-01 10:01` timestamp by filtering in the row that falls between the `measurement_time` and `measurement_end` columns:

measurement_time timestamp	wind_speed int64	temp int64	measurement_end timestamp
2023-12-01 10:00:00	7	-9	2023-12-01 11:00:00

Figure 10.14 – The weather measurement for the time of the ski run

We get the weather data from the most recent measurement at the moment the skier took to the ski run.

This is all getting messy – and gets progressively more complicated when we wish to use this logic to find the most recent weather measurement for each of our skiers. So, let's use DuckDB's ASOF joins to join on nearby time values:

```
SELECT *
FROM scores AS s ASOF JOIN weather AS w
ON s.score_time >= w.measurement_time;
```

This results in the following output:

score_time timestamp	score int64	measurement_time timestamp	wind_speed int64	temp int64
2023-12-01 10:01:00	8	2023-12-01 10:00:00	7	-9
2023-12-01 10:42:00	9	2023-12-01 10:00:00	7	-9
2023-12-01 11:24:00	4	2023-12-01 11:00:00	19	-8
2023-12-01 14:23:00	7	2023-12-01 14:00:00	10	-2
2023-12-01 15:22:00	6	2023-12-01 15:00:00	10	-2
2023-12-01 15:41:00	9	2023-12-01 15:00:00	10	-2
2023-12-02 10:21:00	9	2023-12-02 10:00:00	8	-4
2023-12-02 11:01:00	5	2023-12-02 11:00:00	8	-2
2023-12-02 12:23:00	7	2023-12-02 12:00:00	10	-1
2023-12-02 13:06:00	8	2023-12-02 13:00:00	10	0

Figure 10.15 – The weather measurement for the time of the ski run

`ASOF JOIN` matches the closest (or most recent) value in the `weather` table to each row in the `scores` table based on a `score_time` value that is greater than or equal to `measurement_time` in the `weather` table.

This is pretty neat – we have quickly found the weather conditions that each of our skiers faced during their ski run. The `ASOF` join condition is an elegant way of finding the value of an attribute that's relevant at a specific point in time.

Before we leave this section, let's introduce a very simple visualization that's available in DuckDB to help us appreciate our data.

The DuckDB `bar` function draws a simple graphical bar where the width is proportional to the value within a given range between a specified minimum and maximum value. This will help us see the times when skiers face a very high wind:

```
SELECT s.*,
    bar(w.wind_speed, 0, 20, 20) AS wind_bar_plot,
    w.wind_speed
FROM scores AS s ASOF JOIN weather AS w
ON s.score_time >= w.measurement_time;
```

The `bar(wind_speed, 0, 20, 20)` function plots a horizontal black bar where `wind_speed` is drawn proportional to the expected range between 0 and 20 – and the column is 20 characters wide:

score_time timestamp	score int64	wind_bar_plot varchar	wind_speed int64
2023-12-01 10:01:00	8		7
2023-12-01 10:42:00	9		7
2023-12-01 11:24:00	4		19
2023-12-01 14:23:00	7		10
2023-12-01 15:22:00	6		10
2023-12-01 15:41:00	9		10
2023-12-02 10:21:00	9		8
2023-12-02 11:01:00	5		8
2023-12-02 12:23:00	7		10
2023-12-02 13:06:00	8		10

Figure 10.16 – A bar graph within our query

From this, we can see the times a skier faced a strong wind. Perhaps this can explain why our third skier had such a poor result. Regardless, the bar function offers an effective way to quickly visualize results within a DuckDB query. We will close this chapter by looking at techniques for querying hierarchical data in DuckDB.

Recursive queries and macros

DuckDB supports recursive queries, which are useful for working with hierarchical data, such as organizational charts, product categories, and family trees.

Let's look at how to traverse the hierarchy of wine types and retrieve information about our wine varieties efficiently.

Wine classification hierarchy

Let's start with a table of wine with the association between the wine variety hierarchy expressed as a subclass column:

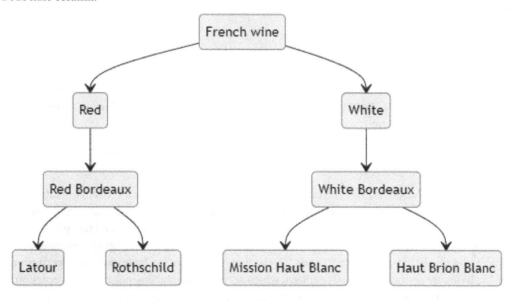

Figure 10.17 – Simplified wine hierarchy

We can create and load the wines table like this:

```
CREATE OR REPLACE TABLE wines AS
SELECT *
FROM read_csv('wines.csv');
```

Now, let's look at the contents of the `wines` table:

```
SELECT *
FROM wines
ORDER BY wine_id;
```

This query returns a small collection of wines and wine varieties:

| wine_id | wine_name | sub_class_of |
int64	varchar	int64
1	French wine	
2	Red	1
3	White	1
4	White Bordeaux	3
5	Mission Haut Blanc	4
6	Haut Brion Blanc	4
7	Red Bordeaux	2
8	Latour	7
9	Rothschild	7

If we look at the `Rothschild` wine, we'll see that it is a subclass of `wine_id` 7, the `Red Bordeaux` classification. In turn, `Red Bordeaux` is a subclass of `wine_id` 2, `Red`, which is a subclass of `wine_id` 1, `French wine`.

We can describe this as a hierarchical relationship, where Rothschild → Red Bordeaux → Red → French wine.

This style of hierarchical traversal is well suited to solving problems using a recursive solution. A recursive pattern is a function that calls upon itself, breaking a large problem down into smaller parts. DuckDB can solve this style of nested problem by using a recursive CTE. By using this pattern, we will be able to define a recursive CTE query that can produce our target result set by recursing over our hierarchical relationships. A recursive CTE consists of two parts: a **base case** and a **recursive case**. The base case defines the initial result set, which serves as the starting point for the recursion. The recursive case defines how to join the CTE to the result of the subsequent invocation. Subsequent applications of the returned case generate new rows until a termination condition is met, which is the point where the recursive case returns no additional rows.

We will be using a recursive CTE to walk through a hierarchy, but it's worth noting that they allow for all manner of powerful data traversal. Some other applications that recursive CTEs can be used for are finding routes between cities, evaluating network conditions, or parsing elaborate paths between nested data.

Let's create a recursive query so that we can navigate our wine collection:

```
WITH RECURSIVE wine_hierarchy(wine_id, start_with, wine_path) AS(
    SELECT wine_id, wine_name, [wine_name] AS wine_path
    FROM wines
    WHERE sub_class_of IS NULL
    UNION ALL
    SELECT wines.wine_id, wines.wine_name,
        list_prepend(
            wines.wine_name, wine_hierarchy.wine_path)
    FROM wines, wine_hierarchy
    WHERE wines.sub_class_of = wine_hierarchy.wine_id
)
SELECT wine_path
FROM wine_hierarchy
WHERE start_with = 'Rothschild';
```

This returns a list of all the wines that are ancestors of Rothschild in the hierarchy, along with their full `wine_path`:

```
|                         wine_path                      |
|                         varchar[]                      |
|--------------------------------------------------------|
| [Rothschild, Red Bordeaux, Red, French wine]           |
```

We started with the `Rothschild` wine, found its ancestor, `Red Bordeaux`, and traversed up the wine hierarch to `Red`, followed by the parent class, `French wine`.

DuckDB's support for recursive queries allows us to traverse the data stored in a hierarchy format efficiently. Here, we have only scratched the surface of what's possible with recursive SQL queries. If you are interested in a deeper understanding of the theory and applications of recursive SQL, there are some great online resources, such as `https://builtin.com/data-science/recursive-sql`.

Before we close this chapter, let's look at macros as another way to structure our queries.

Macros

A DuckDB macro can be used to describe a helper routine, which can simplify code if we wish to encapsulate a shared calculation. A scaler macro can be used in a query, and it allows parameters to be passed into a calculation.

Let's see how a macro can be created to calculate the unit cost of wines:

```
CREATE OR REPLACE MACRO unit_price(price, capacity)
    AS round(price / capacity, 3);
```

The calculation for our unit_price macro simply takes the price of the wine (in USD) and divides it by the capacity (in milliliters). The cost per unit is returned and rounded to three decimal places.

To test out our new function, we'll create a wine_prices table and populate it with some demonstration wines, prices, and bottle capacities:

```
CREATE OR REPLACE TABLE wine_prices AS
SELECT *
FROM read_csv('wine_prices.csv');
```

With the table created, let's look at the data that was loaded in the wine_prices table:

```
SELECT wine_name, price, capacity_ml
FROM wine_prices;
```

This query shows three rows:

wine_name varchar	price double	capacity_ml int64
French Rose	16.5	750
French Pinot Noir	17.0	750
Rose	18.4	1000

We can now use our unit_price macro in a query:

```
SELECT wine_name,
    price,
    capacity_ml,
    unit_price(price, capacity_ml) AS price_ml
FROM wine_prices;
```

This returns the results of the macro scaler function, which displays the calculated unit cost:

wine_name varchar	price double	capacity_ml int64	price_ml double
French Rose	16.5	750	0.022
French Pinot Noir	17.0	750	0.023
Rose	18.4	1000	0.018

The calculation for `unit_price` has been applied to each column. Macros can be considered a useful mechanism for encapsulation logic that can be shared.

Summary

This chapter has offered a range of tips for effective DuckDB usage. Here, we shared time-saving shortcuts for querying and adding data to a DuckDB database. We also introduced various approaches to joining tables, including positional and temporal joins, which provide interesting opportunities for analysis with DuckDB. We also learned how to use recursive queries and macros to handle hierarchical data structures.

As you continue your journey with DuckDB, we encourage you to explore further and experiment with the effective features and functionalities enhancements being added to the project.

With our understanding of effective DuckDB usage, we're in a great position to move on to the next chapter, where we'll dive into hands-on data analysis.

11

Hands-On Exploratory Data Analysis with DuckDB

DuckDB is particularly well-suited to data analysis workflows due to its versatility and highly optimized performance, allowing practitioners to scale data analysis workflows beyond what they would be otherwise able to achieve on their local machine. Previously, we have been focusing more on covering core DuckDB concepts and features, with a bit of data analysis thrown in as examples. In this chapter, we'll be putting the data analysis workflow first by taking what we've learned about using DuckDB and using these foundations to perform some hands-on exploratory analysis of a dataset. This will allow us to explore different approaches for performing effective data analysis with DuckDB.

More specifically, in this chapter, we'll cover the following topics:

- Loading our dataset from a CSV file and applying some data cleaning steps, before writing it to a DuckDB database for our analysis
- Using the JupySQL library for convenient SQL queries in Jupyter Notebooks
- Using the Plotly data visualization library to create interactive data visualizations
- Using DuckDB and Plotly together to perform exploratory data analysis over a public dataset of pedestrian traffic counts through Melbourne CBD

By the end of this chapter, you will be in a position to assemble workflows for performing effective data analysis with DuckDB that works for your particular needs.

Technical requirements

In this chapter, we'll be working with Python inside a Jupyter Notebook environment. You will find the code examples covered in this chapter in the `chapter_11` folder in this book's GitHub repository, which can be found at `https://github.com/PacktPublishing/Getting-Started-with-DuckDB/tree/main/chapter_11`. To work through the exercises here, if you haven't already, you'll need to clone this repository to the machine you'll be using.

Setting up the environment

To be able to run the code in this chapter, you'll need to have a Python virtual environment set up with the necessary dependencies installed. You'll also need a Jupyter-Notebook-compatible IDE setup. If you have not done so already, please follow the instructions in the *Technical requirements* section of *Chapter 7* before jumping into the code in this chapter.

Obtaining the dataset

In this chapter, we'll be working with one of the pedestrian counting-system datasets that was produced and made public by the city of Melbourne, which contains hourly pedestrian counts from pedestrian sensors located in and around the Melbourne Central business district. We'll be working with a historical snapshot of this dataset covering 2009 to 2022.

To download this snapshot, visit the dataset's home page (`https://data.melbourne.vic.gov.au/explore/dataset/pedestrian-counting-system-monthly-counts-per-hour`) and locate the ZIP file containing the 2009 to 2022 archive. Note that this download can be found as an attachment and is a different download from the current version of this dataset, which has a different schema from the archived snapshot that we'll be working with.

Once you've extracted the CSV file from the ZIP archive and placed it inside the `chapter_11` directory, you're good to go. Note that the CSV file in the archive we'll be working with is named `Pedestrian_Counting_System_Monthly_counts_per_hour_may_2009_to_14_dec_2022.csv`. However, for the exercises in this chapter, we've renamed it to the shorter `pedestrian_records_2009-2022.csv`.

Preparing the pedestrian traffic dataset for analysis

Before we can start analyzing the dataset, we need to load it into DuckDB in a shape that supports the kinds of analytical queries that will enable us to explore our dataset effectively. Our plan of attack will be to establish the steps required for parsing and transforming the CSV-backed dataset into a useful schema of columns and data types. After, we'll load this data into a table in a persistent on-disk DuckDB database, which will enable ongoing analysis across working sessions.

As we discussed in *Chapter 8*, when working with DuckDB in Python, we have a choice between using the Relational API or the DB-API to work with DuckDB. The DB-API tends to be better suited when building applications as it promotes interoperability across data processing tools, whereas the Relational API offers a richer interface and features that are designed to enable effective data analysis; so, we will be working with the Relational API for our data analysis.

Reading and transforming the CSV-based dataset

The first step in our analysis will be to identify the data loading and transformation steps we'll use to get the data ready for performing our set of exploratory analyses. We'll do this first step using the DuckDB Python client's default database before we persist our transformation in a table in an on-disk database. As you may recall from *Chapter 7*, the default database is the in-memory database that's automatically created when you import the duckdb module. This is used when you call methods associated with DuckDB connection objects against the duckdb module, such as duckdb.sql() and duckdb.read_csv().

Let's start by reading the CSV using the read_csv() method. As this method is part of the Relational API, it will return a relation object, which we'll assign to a variable:

```
import duckdb
records = duckdb.read_csv("pedestrian_records_2009-2022.csv")
records.show(max_width=200)
```

Note that we've explicitly called the relation's show() method, rather than relying on the notebook to call it implicitly for us. This is so that we can set max_width to a higher value, to prevent columns from being skipped and displayed. The result of displaying the relation we created is as follows:

ID int64	Date_Time varchar	Year int64	Month varchar	Mdate int64	Day varchar	Time int64	Sensor_ID int64	Sensor_Name varchar	Hourly_Counts int64
2887628	November 01, 2019 05:00:00 PM	2019	November	1	Friday	17	34	Flinders St-Spark La	300
2887629	November 01, 2019 05:00:00 PM	2019	November	1	Friday	17	39	Alfred Place	604
2887630	November 01, 2019 05:00:00 PM	2019	November	1	Friday	17	37	Lygon St (East)	216
2887631	November 01, 2019 05:00:00 PM	2019	November	1	Friday	17	40	Lonsdale St-Spring St (West)	627
2887632	November 01, 2019 05:00:00 PM	2019	November	1	Friday	17	36	Queen St (West)	774
2887633	November 01, 2019 05:00:00 PM	2019	November	1	Friday	17	29	St Kilda Rd-Alexandra Gardens	644
2887634	November 01, 2019 05:00:00 PM	2019	November	1	Friday	17	42	Grattan St-Swanston St (West)	453
2887635	November 01, 2019 05:00:00 PM	2019	November	1	Friday	17	43	Monash Rd-Swanston St (West)	387
2887636	November 01, 2019 05:00:00 PM	2019	November	1	Friday	17	44	Tin Alley-Swanston St (West)	27
2887637	November 01, 2019 05:00:00 PM	2019	November	1	Friday	17	35	Southbank	2691
·	·	·	·	·	·	·	·		·
·	·	·	·	·	·	·	·		·
·	·	·	·	·	·	·	·		·
2897597	November 09, 2019 10:00:00 AM	2019	November	9	Saturday	10	27	QV Market-Peel St	371
2897598	November 09, 2019 10:00:00 AM	2019	November	9	Saturday	10	28	The Arts Centre	1188
2897599	November 09, 2019 10:00:00 AM	2019	November	9	Saturday	10	31	Lygon St (West)	229
2897600	November 09, 2019 10:00:00 AM	2019	November	9	Saturday	10	30	Lonsdale St (South)	391
2897601	November 09, 2019 10:00:00 AM	2019	November	9	Saturday	10	34	Flinders St-Spark La	111
2897602	November 09, 2019 10:00:00 AM	2019	November	9	Saturday	10	37	Lygon St (East)	133
2897603	November 09, 2019 10:00:00 AM	2019	November	9	Saturday	10	40	Lonsdale St-Spring St (West)	154
2897604	November 09, 2019 10:00:00 AM	2019	November	9	Saturday	10	36	Queen St (West)	249
2897605	November 09, 2019 10:00:00 AM	2019	November	9	Saturday	10	29	St Kilda Rd-Alexandra Gardens	448
2897606	November 09, 2019 10:00:00 AM	2019	November	9	Saturday	10	42	Grattan St-Swanston St (West)	288

```
? rows (>9999 rows, 20 shown)                                                              10 columns
```

Figure 11.1 – The relation that's produced from loading the pedestrian dataset without any cleaning

As we can see, each record in this dataset contains the number of pedestrian counts passing through a given sensor for a specific hour, along with other information about the hourly reading, such as the sensor name and timestamp of the record, as well as date and time components extracted from the timestamp. Remember that the result sets of relations are loaded lazily, so this only provides a preview of the first 10,000 records that have been loaded. Shortly, we'll see that there are more records than this in the dataset.

The primary data issue that you might have already noticed from this relation is that the Date_Time field is of the VARCHAR type as the CSV reader hasn't been able to automatically detect and parse the non-standard timestamp format of the field. For us to be able to perform some time series analysis and visualizations, we'll need to load this column as a TIMESTAMP type. Let's address this by providing appropriate parameters to the read_csv() method:

```
records = duckdb.read_csv(
    "pedestrian_records_2009-2022.csv",
    dtype={"Date_Time": "TIMESTAMP"},
    timestamp_format="%B %d, %Y %H:%M:%S %p",
)
```

To parse the Date_Time field as TIMESTAMP, we needed to use the dtype parameter of the read_csv() method to indicate the field's type and we also needed to use the timestamp_format parameter to instruct the CSV reader how to parse the string value into a timestamp. For the full reference on format specifiers that can be used when parsing dates and timestamps, see the DuckDB documentation: http://duckdb.org/docs/sql/functions/dateformat.

Let's check if our data cleaning has worked by looking at the first five records of the records relation:

```
records.limit(5).show(max_width=200)
```

This gives us the following relation:

ID int64	Date_Time timestamp	Year int64	Month varchar	Mdate int64	Day varchar	Time int64	Sensor_ID int64	Sensor_Name varchar	Hourly_Counts int64
2887628	2019-11-01 17:00:00	2019	November	1	Friday	17	34	Flinders St-Spark La	300
2887629	2019-11-01 17:00:00	2019	November	1	Friday	17	39	Alfred Place	604
2887630	2019-11-01 17:00:00	2019	November	1	Friday	17	37	Lygon St (East)	216
2887631	2019-11-01 17:00:00	2019	November	1	Friday	17	40	Lonsdale St-Spring St (West)	627
2887632	2019-11-01 17:00:00	2019	November	1	Friday	17	36	Queen St (West)	774

Figure 11.2 – The relation that's produced after parsing the timestamp column correctly

We can see that the Date_Time column now has the correct type of TIMESTAMP.

Using ENUM types to optimize the data loading process

A further enhancement we could make to our data loading process would be to store the Month and Day columns as custom ENUM types. ENUM types provide a mapping from numerical values to the unique values seen in a column, with each value being encoded by the corresponding integer for its value. In our case, we could define two custom ENUM types: one for month names and one for the days of the week, using them as types for the Month and Day columns, respectively. This is particularly beneficial when we're dealing with string columns containing categorical values with low cardinality (that is, low numbers of distinct values), as is the case here, since the resulting ENUM columns only contain more efficient numerical values. This can result in a considerable reduction in storage space, as well as faster query performance over these columns. For more information about creating and using ENUM types, consult the DuckDB documentation: `https://duckdb.org/docs/sql/data_types/enum.html`.

Before loading our dataset into an on-disk database for analysis, we can also consider whether there are any data transformations we may want to apply. For example, we could drop any columns that we know we won't need. The additional date and time fields beyond the timestamp of the record are likely to be useful for us, and the `Sensor_ID` column can be used to link this dataset with a different dataset published by the city of Melbourne that contains information about each sensor, such as its geographic coordinates, so we should keep all these. The `ID` field, on the other hand, is not found in any other dataset concerning the pedestrian counting system, so we can drop this. Given that we're going to be performing a range of time series analyses, we'll almost certainly need these records to be ordered by timestamp, so the other transformation we'll perform involves sorting the records by the `Date_Time` column. Let's apply these transformations to our relation and inspect the result:

```
transformed = records.select("* EXCLUDE ID").sort("Date_Time")
```

In this transformation, we used the `select()` method of the records relation object, passing it a SQL expression as a string that selects all columns except ID. Then, we used the resulting relation's `sort()` method, sorting by the `Date_Time` column. Recall from *Chapter 8* that when you're working with the Relational API, relation methods that require expressions as arguments can generally take either SQL expressions as strings or expression objects. Often, the use of SQL expressions is more convenient for interactive data analysis. In the case of the `select()` call, we used the convenient EXCLUDE keyword, which, as we saw in *Chapter 10*, enables a concise column section.

Let's inspect the first five results from our `transformed` relation:

```
transformed.limit(5).show(max_width=200)
```

This produces the following output:

Date_Time timestamp	Year int64	Month varchar	Mdate int64	Day varchar	Time int64	Sensor_ID int64	Sensor_Name varchar	Hourly_Counts int64
2009-05-01 00:00:00	2009	May	1	Friday	0	1	Bourke Street Mall (North)	53
2009-05-01 00:00:00	2009	May	1	Friday	0	2	Bourke Street Mall (South)	52
2009-05-01 00:00:00	2009	May	1	Friday	0	4	Town Hall (West)	209
2009-05-01 00:00:00	2009	May	1	Friday	0	5	Princes Bridge	157
2009-05-01 00:00:00	2009	May	1	Friday	0	6	Flinders Street Station Underpass	139

Figure 11.3 – The relation that's produced after applying our data transformations

Both the schema and shape of our data are now looking pretty good for performing our analysis. Of course, the nature of data analysis is iterative, and we may find that we need to come back to tune our data processing steps after working with the dataset. But for now, we're ready to load this dataset into a persistent database.

Loading the prepared dataset into a database table

Having established a process to ingest our CSV, we'll switch from using the default in-memory database to using a new on-disk database where we will load and persist our cleaned dataset as a table. Doing this gives us the advantage of only having to process our data once, so we won't have to wait for the CSV parsing to happen every time we load our notebook. This is a particularly useful pattern for larger and more complex datasets, where the data loading process could be computationally intense and take some time. Another advantage of this approach is that it provides separation of concerns regarding data loading and data consumption, allowing multiple notebooks that perform different types of analyses to consume from the database without having to be concerned with data ingestion logic.

Given that we need to write to a specific database, we'll need to create a connection to a new on-disk database. Since we must close this connection after we're done writing our table to ensure our new database is updated safely, we'll use the new connection as a context manager, an approach to managing database connections that we discussed in *Chapter 7*. Inside the context manager's with block, we'll put all the data preparation steps, as well as create our new table, which we'll call pedestrian_counts. This will have the effect of automatically closing the connection after we finish writing to it:

```
with duckdb.connect("pedestrian.duckdb") as conn:
    result = (
        conn.read_csv(
            "pedestrian_records_2009-2022.csv",
            dtype={"Date_Time": "TIMESTAMP"},
            timestamp_format="%B %d, %Y %H:%M:%S %p",
        )
        .select("* EXCLUDE ID")
        .sort("Date_Time")
    )
    result.to_table("pedestrian_counts")
```

Let's unpack what we did to load our transformed dataset into a persistent database. We started by defining a `with` block that uses a new on-disk database connection as a context manager, within which we can interact with our new database that will be persisted in the `pedestrian.duckdb` database file. We also combined all our previous data processing steps into a combined chain of Relational API method calls, with the `read_csv()` method creating a relation object that represents the CSV-reading component of the query and the `select()` method of the resulting relation being invoked with the `sort()` method, each adding to the final complete query logic being captured in the `results` variable. We finished by calling the `to_table()` method on our resulting relation, which is when the lazy relation object will be executed, so that it can write our transformed dataset to the `pedestrian_counts` table in the `pedestrian.duckdb` database.

Now that we've loaded our transformed dataset into a persistent database, we're ready to cover some supplementary tools that are going to support our Jupyter-Notebook-based analysis.

Effective data analysis using Jupyter Notebooks

In this section, we're going to briefly cover two open source tools that will assist us in performing our data analysis within a Jupyter Notebook. The first is **JupySQL**, which provides us with a convenient way to run SQL queries in Jupyter notebooks. The second is **Plotly**, a comprehensive data visualization library that produces interactive visualizations with strong support for running inside Jupyter Notebooks. Let's get started.

Convenient SQL queries with JupySQL

JupySQL is an open source Python package for streamlining the process of writing and running SQL queries in Jupyter Notebooks. To understand how it can help our data analysis workflow, let's have a look at how we can use the Relational API to query our `pedestrian_counts` table using SQL. Say we wanted to count the total number of pedestrian readings for the *Melbourne Central* sensor for 2022. We might write the following query and run it using the `sql()` method from the Relational API:

```
conn = duckdb.connect("pedestrian.duckdb")
conn.sql(
    """
    SELECT sum(Hourly_Counts) AS Total_Counts
    FROM pedestrian_counts
    WHERE Year = 2022 AND Sensor_Name = 'Melbourne Central'
    """
)
```

This gives us the following relation with the answer to our question:

Total_counts int128
6897406

Figure 11.4 – The DuckDB relation produced from our SQL query that calculates the
total number of pedestrian readings in 2022 for the Melbourne Central sensor

This approach to writing SQL becomes a bit cumbersome as it means having to do all your SQL query
development inside a Python string. Using a triple-quoted string, as we did here, helps a little, but we
still need to wrap every SQL query in the `sql ()` method call, as well as the triple quotes, which will
slow down our analysis feedback loop. It would be much more ergonomic if we could simply write the
contents of our SQL string directly into a Jupyter Notebook cell and have the SQL query be executed
appropriately and automatically. This is exactly what JupySQL allows us to do.

To use JupySQL in a notebook, we need to configure JupySQL so that it knows which database to
submit our queries to. Here's a complete example of configuring your notebook so that it uses JupySQL
with a new in-memory database:

```
%load_ext sql
conn = duckdb.connect()
%sql conn --alias duckdb
```

The commands prefixed with % characters are not Python code, but rather **magic commands** (or
just *magics*), which are directives that are made available by the IPython kernel – the default Python
kernel used by Jupyter. Magics enable you to interact with and configure the underlying kernel that
runs your Python code inside the notebook. In the preceding example, `%load_ext` is a line-magic
that loads the JupySQL extension into the notebook's kernel, at which point the `%sql` line-magic
tells JupySQL to use the new in-memory database, `conn`, we created, while also giving it an alias
of `duckdb`, which helps us keep track of which connection was used for each query in case we're
working with multiple databases.

In our context, we want to work with the database we created in the previous section; so, let's configure
JupySQL so that it uses a new connection to our on-disk `pedestrian.duckdb` database while
also giving it an appropriate alias:

```
conn = duckdb.connect("pedestrian.duckdb")
%sql conn --alias pedestrian.duckdb
```

Alternative JupySQL-DuckDB configuration

In addition to using native DuckDB connections, there is an alternative way to connect JupySQL to DuckDB that involves connecting via the SQLAlchemy Python library. This enables you to use JupySQL's `%sql` cmd line-magic, which is not supported with DuckDB native connections, and which allows you to gather a range of diagnostic information about tables in your database, as well as provide an interactive widget for creating and managing database connections. See the JupySQL docs for more details: `https://jupysql.ploomber.io/en/latest/tutorials/duckdb-native-sqlalchemy.html`.

By default, JupySQL returns its own lazily-evaluated result-set object, which can then be operated on or converted into a dataframe. However, to conduct exploratory data analysis, rather than having to convert this into a dataframe each time, it would be more convenient to get back a dataframe directly. JupySQL supports conversion to both pandas and Polar dataframes, and for this analysis, we'll use pandas. So, let's configure JupySQL so that it automatically returns a pandas dataframe for each query:

```
%config SqlMagic.autopandas = True
```

An important caveat to be aware of is that by setting this parameter, we will lose the lazy evaluation that is associated with DuckDB's relation objects. All SQL commands that we run with JupySQL will have their query results fully materialized so that they can be converted into pandas dataframes. For our purposes, with the analyses we want to perform on this dataset, this will be fine; however, it is important to know that this does have performance implications, especially when working with large datasets.

With JupySQL configured, we can now run our previous SQL query much more simply through the use of JupySQL's `%%sql` cell-magic. Jupyter cell-magics provide instructions to the IPython kernel to change how the contents of the cell are run. In this case, the `%%sql` cell-magic causes the contents of the cell to be submitted to the configured database as a SQL string for running. Let's see this in action:

```
%%sql
SELECT sum(Hourly_Counts) AS Total_Counts
FROM pedestrian_counts
WHERE Year = 2022 AND Sensor_Name = 'Melbourne Central'
```

Running this cell gives us the following output:

Running query in 'pedestrian.duckdb'

	Total_Counts
0	6897406.0

Figure 11.5 – A pandas dataframe produced from a DuckDB SQL query using JupySQL

We can see the same answer that we saw previously, but this time, it's displayed as a pandas dataframe, which JupySQL automatically converted our query results into. You might have also noticed that our dataframe has a decimal value rather than an integer. We'll come back to addressing this shortly.

Note that after running this cell, we haven't captured this dataframe as a variable; we've simply displayed it. We can access this resulting dataframe through the special _ variable, which the IPython kernel automatically assigns the value of the final expression in the most recently executed cell. To illustrate this, we can use Python's built-in type() function:

```
type(_)
```

This shows us that the _ variable was indeed assigned a pandas dataframe:

```
pandas.core.frame.DataFrame
```

This is a little awkward as we will often want a persistent reference to the resulting dataframe from our queries so that we can use it in subsequent steps of our analysis. Fortunately, JupySQL provides a convenient syntax for assigning the result of our query to a new variable. All we have to do is change our %%sql magic to %%sql query_result <<, which will assign the resulting dataframe to the query_result variable. Let's try this out with a different SQL query that calculates the total counts for each sensor in 2022 and sorts them in descending order:

```
%%sql sensors_2022_df <<
SELECT Sensor_Name, sum(Hourly_Counts)::BIGINT AS Total_Counts
FROM pedestrian_counts
WHERE Year = 2022
GROUP BY Sensor_Name
ORDER BY Total_Counts DESC
```

We don't see any output after executing this cell as our resulting dataframe has been assigned to the sensors_2022_df variable. Let's inspect the first 10 rows of this dataframe to find the top sensors for 2022 by total pedestrian counts:

```
sensors_2022_df.head(10)
```

This gives us the following output:

	Sensor_Name	Total_Counts
0	Flinders La-Swanston St (West)	10492872
1	Southbank	8737282
2	Melbourne Central	6897406
3	Elizabeth St - Flinders St (East) - New footpath	6511465
4	Princes Bridge	6202149
5	State Library - New	6049385
6	Flinders Street Station Underpass	5772514
7	Melbourne Convention Exhibition Centre	5634531
8	Bourke Street Mall (North)	5614610
9	Melbourne Central-Elizabeth St (East)	5380759

Figure 11.6 – A pandas dataframe with the top 10 sensors by total pedestrian counts for 2022

Before we move on, we'll take a moment to comment on the expression in this query that produced the Total_Counts column. You might have noticed the ::BIGINT component. This casts the values of the column to a BIGINT data type, which represents integers using 64 bits. We did this to address the issue we saw earlier, where our dataframe contained a decimal value rather than an integer, as we might have expected. DuckDB's sum aggregate returns 128-bit integers (available as the HUGEINT data type) to prevent overflow errors, which would otherwise occur when the data being summed generates a larger number than what can be represented with the more commonly used INTEGER and BIGINT data types. pandas does not support 128-bit integers, however, so it converts them into its float64 data type, which can store a larger range of numbers, but is not an integer. This is why we saw a decimal value in the dataframe containing the total number of pedestrian readings for the *Melbourne Central* sensor in 2022. Casting the result of sum to BIGINT, which pandas can convert into its int64 data type, ensures that we wind up with integers in our dataframe. We can confirm that this has worked by inspecting the dtypes attribute of our dataframe:

```
sensors_2022_df.dtypes
```

This gives us:

```
Sensor_Name     object
Total_Counts     int64
dtype: object
```

In addition to confirming that the `Total Counts` column of our dataframe has an `int64` data type, we can also see that the `Sensor Name` column contains the `dtype` object. While this can be a little misleading at first, this is pandas' default way of representing string values and is therefore appropriate for this column.

The dataframe we produced with our query does a decent job of showing us the top 10 sensors ranked by traffic. However, if we want to understand the traffic behind these sensors in more depth, we'll probably want to start visualizing our data. In the next section, we'll see how we can use Plotly to make interactive data visualizations directly in our notebook.

Interactive visualizations with Plotly

Plotly is an open source Python package for producing data visualizations. Its ability to make interactive browser-based data visualizations makes it particularly useful in exploratory data analysis workflows, where interactive charts serve as valuable tools. This will be a very quick overview of working with Plotly so that we can use it within our data analysis. For a more comprehensive guide, see the Plotly Python client documentation: `https://plotly.com/python`.

The Plotly Python client is a wrapper around the underlying Plotly.js JavaScript library. When working with the Python client, we are essentially building chart configuration objects that are fed into Plotly.js for rendering in a browser. Plotly refers to these objects that encode our visualization's properties as **figures**. To make these figures, you can either build them up using lower-level components of the **Plotly Python API** or you can use **Plotly Express**, a higher-level interface for rapidly creating Plotly `Figure` objects which is also included in the Plotly Python package. In general, a good heuristic is to use Plotly Express where possible, falling back to the lower-level interface only when you need more customization than is possible using Plotly Express alone. This is what we'll be doing when we conduct data analysis in this chapter.

Plotly Express is primarily designed around creating visualizations from dataframes, supporting both pandas and Polars, with non-dataframe input also being supported. The workflow we'll use is to query DuckDB using JupySQL, and then feed the resulting dataframe into Plotly Express. The best way to work with Plotly Express is to import the `plotly.express` module under the more convenient name of `px`. Once we've done this, we can use it to call specific visualization functions that are needed for our analysis:

```
import plotly.express as px
```

With the Plotly Express module loaded, let's see it in action by using the `px.bar()` function to create a bar chart of the top 10 sensors by traffic in 2022 while using the `sensors_2022_df` dataframe we created in the previous section:

```
figure = px.bar(
    sensors_2022_df.head(10),
    x="Sensor_Name",
```

```
        y="Total_Counts",
        height=500,
        title="Top 10 sensors by traffic for 2022",
    )
    figure
```

By ending the cell with the `figure` variable that contains the `Figure` object that's returned from the `px.bar()` function, most Jupyter Notebook environments should automatically render this as a Plotly visualization. However, in some cases, you may need to explicitly call `figure.show()`. Once displayed, our figure will look like this:

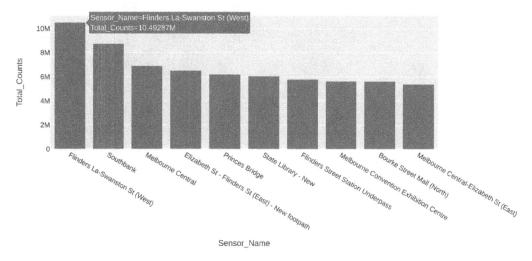

Figure 11.7 – A Plotly bar chart showing the top 10 sensors by total pedestrian counts for 2022

What we've done here is tell Plotly to make a bar chart using the data in the dataframe that was supplied in the first argument, after which we've provided configuration about the properties of the chart to the `px.bar()` function using keyword arguments. The x and y parameters indicate which columns in the dataframe should be used to feed data into the *x* and *y* axes of the chart. We also used the `height` parameter to configure the height of the chart in our notebook so that it's a little larger than it would otherwise be rendered at, as well as using the `title` parameter to provide a title for our chart. The resulting bar chart allows us to compare the relative sizes of the top 10 sensors much more easily. In *Figure 11.7*, we've captured an interactive feature of Plotly's bar charts, which is that hovering over bars will result in a popup that displays the full *x* and *y* values for that data point. Here, we can see that the most trafficked sensor, *Flinders La-Swanston St (West)*, had 10.49 million pedestrian counts.

> **Title your plots!**
>
> As a matter of practice, it's a good habit to ensure all your visualizations have titles. While it may feel like a chore when you just want to get on with producing a plot, you will thank yourself upon coming back to your notebook later when your analysis is no longer fresh in your head. Importantly, this will also make your notebook a much more accessible artifact for anyone else working with or reviewing your notebook.

Let's make one more visualization with Plotly Express. This time, we'll visualize the number of active sensors across each year using a line plot. Plotly provides this functionality through the px.line() function. First, we'll need to query our database to get the distinct number of sensor names seen across each year, which we can do with the following DuckDB query:

```
%%sql sensor_years_df <<
SELECT Year, COUNT(DISTINCT Sensor_Name) AS Total_Sensors
FROM pedestrian_counts
GROUP BY Year
ORDER BY Year
```

After running this, `sensor_years_df` will contain a dataframe of year and sensor number columns:

```
sensor_years_df.head(5)
```

When we peek at the first five rows, we'll see the following dataframe:

	Year	Total_Sensors
0	2009	18
1	2010	18
2	2011	18
3	2012	18
4	2013	32

Figure 11.8 – A pandas dataframe showing the first 5 years from the pedestrian-counts dataset and the number of active sensors for those years

To create a figure object that specifies our desired line plot, we'll call the px.line() function and pass it to our new dataframe, as well as specify the target columns for the *x* and *y* axes. Additionally, we'll use the marker parameter to configure the line plot so that it includes a marker for each data point, without which our plot would simply be a plain line:

```
figure = px.line(
    sensor_years_df,
    x="Year",
    y="Total_Sensors",
    markers=True,
    height=500,
    title="Total number of active sensors by year"
)
figure
```

When evaluated, Plotly produces the following visualization from our figure:

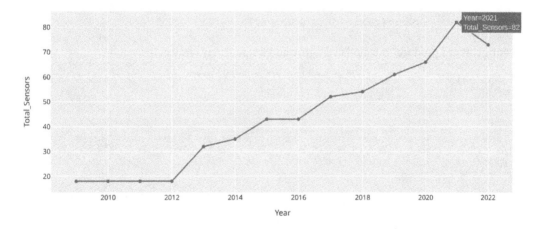

Figure 11.9 – A Plotly line chart showing the number of active sensors by year

From this plot, we can see that there is a general trend for the number of active sensors to grow over the years, except for between 2021 and 2022, when the number dropped slightly. Hovering over these points shows a drop from 82 sensors to 73.

Sometimes, you may want to customize your visualizations in ways that Plotly Express doesn't allow through its more concise interface. In these situations, we can fall back to Plotly's lower-level figure API by updating the properties of the figure object that's returned from Plotly Express functions. For example, in the chart shown in *Figure 11.9*, we might want every year on the x-axis to have a label, rather than every other. We also might want to align the title so that it's in the center of the plot, rather than to the left. A general-purpose way to update layout properties such as these is to call the `Figure.update_figure()` method (for an existing `Figure` object) and pass its parameters with the new properties. Let's use this approach to make a new figure with those changes we identified:

```
figure.update_layout(xaxis={"dtick": 1}, title={"x": 0.5})
```

Here, we've passed in values for the `xaxis` and `title` parameters, which take dictionaries of more fine-grained attributes that specify granular properties of the x-axis and the title, respectively. The `dtick` parameter indicates the number of steps between each tick on the x-axis, and x indicates a number from 0 to 1 that indicates the relative horizontal position to place the title. Plotly also provides a more convenient way of specifying nested properties through its "magic underscore" notation, which allows you to configure nested figure properties by joining property levels together using a _ character. Using this notation, we can rewrite the preceding line so that it's much simpler:

```
figure.update_layout(xaxis_dtick=1, title_x=0.5)
```

Evaluating this cell produces our updated figure and immediately renders it:

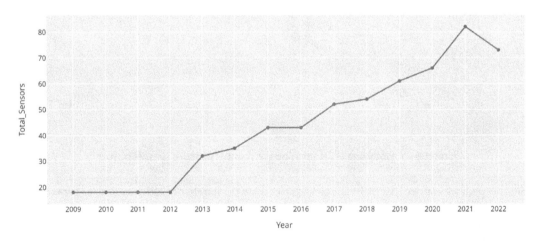

Figure 11.10 – A Plotly line chart showing the same chart from Figure 11.9, with its visual properties changed to make the title centered and labels occur on every x-axis tick

In this short introduction to making visualizations with Plotly, we've focused on two Plotly Express functions, as well as Plotly chart customizations via a handful of figure-layout properties; however, this is by no means a comprehensive overview of using the library. Two resources we'd recommend consulting if you want to further explore using Plotly in Python are as follows:

- The Plotly Express guide: `https://plotly.com/python/plotly-express`
- The Plotly Python Figure reference: `https://plotly.com/python/reference/index`

To recap, the process we adopted was to use the higher-level Plotly Express interface to generate visualizations concisely, after which we tweaked the resulting figure object by updating its layout properties. This is an effective pattern when making Plotly visualizations, which we'll make use of in our analyses in the following sections. Let's jump into it.

Analyzing pedestrian traffic through Melbourne CBD

Now that we've prepared our dataset and covered the tools we're going to use, let's jump into some analysis of the Melbourne pedestrian counting dataset. We'll continue to use DuckDB with JupySQL to query our `pedestrian_counts` table, and Plotly to make visualizations. Note that we won't always display the dataframe showing the results of our query. As you're working through the examples, we encourage you to inspect the contents of the dataframes yourself.

Visualizing total pedestrian counts over time

To start with, let's get a sense of how the total number of pedestrian counts registered by the sensor network has changed over the years in the dataset. We'll start by querying the `pedestrian_counts` table to get the sum of all counts within each year in the dataset:

```
%%sql year_counts_df <<
SELECT Year, sum(Hourly_Counts)::BIGINT AS Total_Counts
FROM pedestrian_counts
GROUP BY Year
ORDER BY Year
```

This gives us a dataframe that contains the `Year` and `Total_Counts` columns, with each record indicating the total number of pedestrian counts for each year. Now, let's use Plotly Express to create a line plot of this by using `px.line()`:

```
px.line(
    year_counts_df,
    x="Year",
    y="Total_Counts",
    markers=True,
    height=500,
```

```
        title="Total pedestrian counts by year",
)
```

Note that we didn't assign the result of the px.line() function to a variable as all we want to do is view the result. Since it's the only expression in this cell, our notebook will still render the resulting Figure object, which will look like this:

Yearly traffic across all sensors

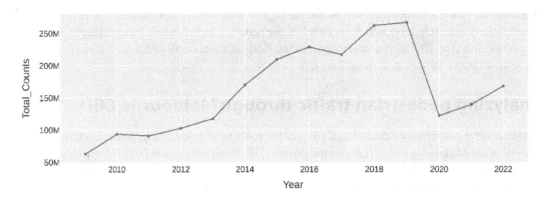

Figure 11.11 – A Plotly line chart showing the total pedestrian counts
recorded across the entire pedestrian counting system by year

This plot shows a steady increase over the years in the number of total pedestrians recorded moving across Melbourne city, with a rather pronounced crash in 2020, which is almost certainly due to the impact that COVID-19 lockdown restrictions had on the number of people walking through the city. However, we need to be careful about the conclusions we draw from this plot. As we saw in the previous section, the number of active sensors in the system has increased over time, which will result in some pedestrians being counted more times over the same route than in previous years.

To address the confounding impact of the changing number of sensors over the years, let's filter down our dataset to only records from sensors that have readings across every year, before repeating the same analysis. We'll do this in two steps:

1. Extract the names of sensors that have readings for every year in the dataset, storing the result in a table called common_sensors.

2. Repeat our sum aggregation of pedestrian counts by year while also filtering our rows with sensor names that don't occur in the common_sensors table.

We can achieve the first step by grouping our dataset by sensor name and selecting only sensor names whose group of records includes at least one record for each year that our dataset covers. We know that our dataset includes data from 2009 to 2022, covering 14 years. Performing a simple SQL query against our dataset confirms that this is the distinct number of years seen in our dataset:

```
%sql SELECT count(DISTINCT Year) FROM pedestrian_counts
```

Note that we've used JupySQL's `%sql` line magic (as opposed to the `%%sql` cell magic) as this is a simple one-line query. Running this gives us the following output:

Running query in 'pedestrian.duckdb'

count(DISTINCT "Year")

0   14

Figure 11.12 – The number of distinct years in our dataset

We'll leverage this observation that each sensor will need to have 14 distinct `Year` values across its readings for it to be considered. Here's the query we'll use to retrieve these corresponding `Sensor_Name` values. We've also used this to populate the `common_sensors` table:

```
%%sql
CREATE OR REPLACE TABLE common_sensors AS
SELECT Sensor_Name
FROM pedestrian_counts
GROUP BY Sensor_Name
HAVING COUNT(DISTINCT Year) = 14;
```

Evaluating this SQL statement results in our query being executed and its results being populated into the new `common_sensors` table. Ordinarily, when issuing a `CREATE TABLE` statement in DuckDB, no results are returned. When using JupySQL to submit database-modifying SQL, however, it returns the number of rows that were inserted into the database. This means we'll see the following dataframe as output:

Running query in 'pedestrian.duckdb'

Count

0   15

Figure 11.13 – The number of rows that were inserted into the newly created common_sensors table

Let's unpack the query we used to populate the `common_sensors` table. To get the distinct sensor names in the dataset, we used GROUP BY on the `Sensor_Name` column, while also selecting this column. We also needed to filter these down to only those sensor names with records for every year in the dataset. The HAVING clause is exactly what we need here as it allows us to filter grouped results formed by a GROUP BY clause. This differs from the WHERE clause, which filters rows before any groupings are made. The filter expression we gave the HAVING clause ensures that a group is only included in our results if the distinct number of Year values from the group is the same as the distinct number of Year values seen across the entire dataset.

Now that we know which sensors were active across all years in the dataset, we can perform the next step. Here, we'll adapt our previous query so that it sums the total number of pedestrian counts for each year. However, this time, we'll restrict pedestrian counts to only those coming from sensors in the `common_sensors` table:

```
%%sql year_counts_filtered_df <<
SELECT Year, sum(Hourly_Counts)::BIGINT AS Total_Counts
FROM pedestrian_counts
WHERE Sensor_Name IN (FROM common_sensors)
GROUP BY Year
ORDER BY Year
```

This is the same query we used previously to return the yearly counts but with an added WHERE clause to filter out the sensors we want to exclude. The filter expression does this using the IN operator, checking to see if the `Sensor_Name` value of each row is present in the single column that's returned from the subquery, which simply selects every row in the `common_sensors` table. Also, note that we've taken advantage of DuckDB's convenient feature of being able to omit the SELECT clause in a query when selecting every column.

Let's feed the dataframe this produced into the same `px.line()` call as before:

```
px.line(
    year_counts_filtered_df,
    x="Year",
    y="Total_Counts",
    markers=True,
    height=500,
    title="Yearly traffic for sensors active all years",
)
```

This produces the following plot:

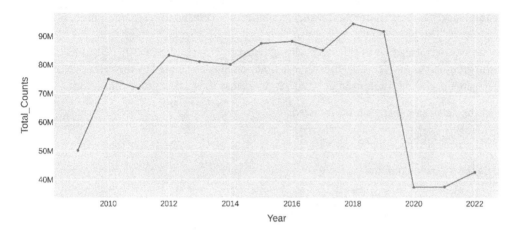

Yearly traffic for sensors active all years

Figure 11.14 – A Plotly line chart showing the total pedestrian counts by year
across only sensors that were active for every year in the dataset

Comparing this with the previous plot, we can see that the trend regarding the increasing number
of pedestrian counts over the years is less pronounced. This is what we would expect to see when
controlling the increasing number of sensors. We can still see the dramatic drop in sensor counts
due to the COVID-19 lockdowns, which, if anything, now looks even more pronounced compared
to the previous years.

To finish up our investigation into pedestrian counts through the city over time, let's drill down a
bit deeper into the impact that the lockdowns had on Melbourne pedestrian traffic by looking at the
month-to-month changes for the years 2019, 2020, and 2021. For this, we'll need a new data extract
from our `pedestrian_counts` table where we group by both years and months and then aggregate
the total counts for each combination of year and month. We'll also filter this down to only readings
from sensors in our `common_sensors` table during the target years:

```
%%sql year_month_counts_df <<
SELECT
    Year,
    Month,
    month(Date_Time) AS Month_Num,
    sum(Hourly_Counts)::BIGINT AS Total_Counts
FROM pedestrian_counts
WHERE Year IN (2019, 2020, 2021)
    AND Sensor_Name in (FROM common_sensors)
```

```
GROUP BY Year, Month, Month_Num
ORDER BY Year, Month_Num
```

An additional step we took in this query was to derive a `Month_Num` column from the `Date_Time` column using DuckDB's `month` function, which contains the corresponding month numbers (1 to 12) for each record. This allowed us to sort our records by both the year and month using the `ORDER BY` clause. It's worth noting that in this case, adding `Month_Num` to the `GROUP BY` clause does not affect the grouping as the `Month_Num` value is functionally determined by the value of `Month`. However, in general, adding columns to the `GROUP BY` clause does affect group formation.

Let's inspect the dataframe that we just produced:

```
year_month_counts_df.head(15)
```

This gives us the following output:

	Year	Month	Month_Num	Total_Counts
0	2019	January	1	7000243
1	2019	February	2	6079776
2	2019	March	3	7740551
3	2019	April	4	8281383
4	2019	May	5	7908534
5	2019	June	6	7323166
6	2019	July	7	7915558
7	2019	August	8	8013224
8	2019	September	9	7436127
9	2019	October	10	7665717
10	2019	November	11	7764366
11	2019	December	12	8547951
12	2020	January	1	7568731
13	2020	February	2	7012371
14	2020	March	3	5000186

Figure 11.15 – A pandas dataframe containing the first 15 records of pedestrian counts, aggregated by the sum of counts for each month and year pair

As you can see, this dataframe contains a record for each year-month pair, with its corresponding total counts, and is sorted by both year and count. To create our line plot, we can call `px.line()` with the x parameter set to `Month`, and the y parameter set to `Total_counts`. We also need to configure the resulting figure object so that it produces multiple lines, one for each year in our data extract. Plotly Express allows us to do this concisely by specifying that the `Year` column in our dataframe should be treated as a categorical variable whose distinct values are used to group the dataset into multiple lines, one for each year. Plotly refers to each data series occurring within a single figure as a **trace**. Plotly provides several ways to split an input dataframe into groups to create multiple traces that can be compared, each of which uses a different visual variable to encode each trace.

Since you may be viewing an electronic version of this book via a color display, or you could be reading a monochrome printed page, we'll use two visual variables to distinguish the different year plots: the color of each line and the shape of the line markers. To do this, we can pass `px.line()` the `color` and `symbol` keyword arguments, both with a value of `Year`. We'll also specify the specific marker shapes to use by passing the `symbol_squence` parameter as a list of symbol-identifier strings. Let's see all of this together:

```
px.line(
    year_month_counts_df,
    x="Month",
    y="Total_Counts",
    color="Year",
    symbol="Year",
    symbol_sequence=["square", "diamond", "circle"],
    markers=True,
    height=500,
    title="Monthly traffic for sensors active 2019-2021",
).update_traces(marker_size=8)
```

The other visual enhancement we made was to improve interpretability by setting the marker size that's used for each trace to a larger value by using the `update_traces()` method. When we evaluate the preceding cell so that we can render the resulting figure, we get the following visualization:

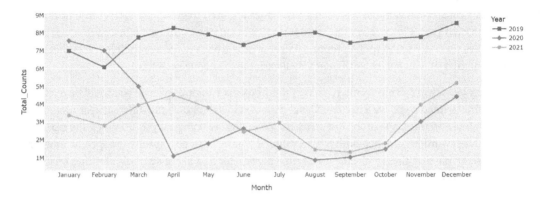

Figure 11.16 – A Plotly line chart containing monthly pedestrian counts,
with a different trace for each year across 2019 to 2021

This visualization allows us to readily compare the month-by-month total counts for each of the target years alongside each other. By hovering over the markers of different lines, we can glean that in 2019, April saw a total of 8.2 million pedestrian counts, which went all the way down to 1.1 million for the same month in 2020, when the initial lockdown was in full effect.

Bear in mind that these total `Total_Count` numbers are only derived from stations that occur in our `common_sensors` table, which are those that are active across every year in the dataset. This means we will have pruned out the sensors that were added after the start of the dataset and were active from 2019 to 2021. To enhance our analysis, which you could try out yourself, we could update the sensor-filtering logic we used so that it's only based on the target years.

Creating time series plots of sensor readings

We'll now move on to a finer level of analysis and explore how the hourly pedestrian traffic changes over a range of days for specific sensors. Let's start by creating a plot of hourly traffic for the *Flinders La-Swanston St (West)* sensor, across all of 2020. First, we'll create an extract of the records in the dataset by getting the `Hourly_Counts` and `Date_Time` columns for only the records that correspond to our target sensor and year:

```
%%sql sensor_2020_df <<
SELECT Hourly_Counts, Date_Time
FROM pedestrian_counts
WHERE Sensor_Name = 'Flinders La-Swanston St (West)' AND Year = 2020
```

Then, we'll use the px.line() function to create a time series plot of all readings for that year:

```
px.line(
    sensor_2020_df,
    y="Hourly_Counts",
    x="Date_Time",
    height=500,
    title="Hourly traffic for Flinders La-Swanston St (West)",
)
```

This produces a figure that renders like this:

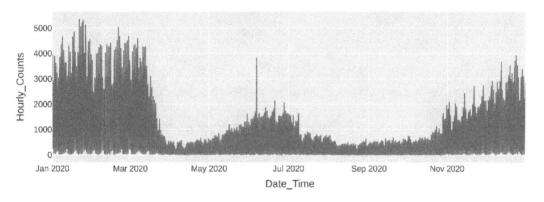

Figure 11.17 – A Plotly line chart visualizing the hourly pedestrian counts
for the Flinders La-Swanston St (West) sensor for 2020

Since we've visualized every hour across the whole year, this is a very dense line plot. It provides a striking visualization of the impact the COVID-19 lockdowns had on pedestrian traffic across this sensor:

- We can see traffic dropping around the middle of March as restrictions are applied, leading up to when the first full lockdown began on March 31

- Traffic starts to increase in mid-May as restrictions are relaxed

- There was a massive spike on June 6, coinciding with a Black Lives Matter protest that occurred close to the central business district

- Traffic drops off again, with a second lockdown being put in place on July 7

- We start seeing traffic increase again from October through to the end of the year, corresponding to the ongoing lifting of restrictions, which started in mid-September

This is a perfect time to take advantage of another of the interactive features that Plotly's visualizations offer. We can zoom in on specific regions of the *x* and *y* axes by left-clicking and dragging over the desired region. Selecting the region from January 19 to 31 gives us the following time series:

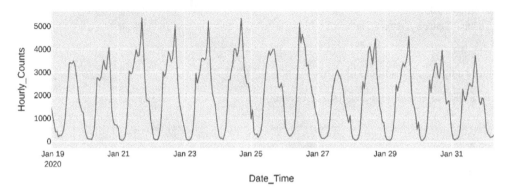

Figure 11.18 – A zoomed-in view of the same time series plot presented in Figure 11.17

We can now easily discern the periodic changes in pedestrian traffic for this sensor that correspond to the daily rhythm of people's activities. To reset the zoom level, you can either double-click anywhere in the figure or click the home icon in the figure's **modebar**, which appears when hovering over Plotly figures.

It might be interesting to compare the daily traffic rhythms across several different sensors to see whether there are variations across different sensors. Let's do this for three sensors, just for September 2019. The DuckDB query we'll need for this is a simple one that filters our records down to the desired sensors and date range:

```
%%sql multi_sensor_df <<
SELECT Sensor_Name, Hourly_Counts, Date_Time
FROM pedestrian_counts
WHERE Sensor_Name IN (
        'Flinders St-Spark La',
        'Bourke Street Mall (North)',
        'Southern Cross Station'
    ) AND Year = 2019 AND Month = 'September'
```

Now, we'll use the `px.line()` function to produce a facet plot, which contains multiple subplots. We can do this with either the `facet_col` or `facet_row` parameters, both of which take a column name and produce subfigures for each distinct value in the column. We'll use the `Sensor_Name` column to produce time series subfigures for each sensor. To support comparison across sensors, these subfigures would be best arranged as horizontal rows on top of each other. We can achieve this by setting the `facet_row` parameter to `Sensor_Name`. However, this happens to result in some rather squished subfigure titles due to our long sensor names. So, instead, we'll achieve the same horizontal layout through stacked columns, which are limited to a single column of subfigures, by using the `facet_col` parameter and setting the `facet_col_wrap` parameter to `1`. Putting this together, our `px.line()` call looks like this:

```
px.line(
    multi_sensor_df,
    y="Hourly_Counts",
    x="Date_Time",
    facet_col="Sensor_Name",
    facet_col_wrap=1,
    title="Hourly pedestrian traffic for December 2019",
    height=800,
).update_layout(yaxis_fixedrange=True)
```

The one layout change we made to the entire figure was to the `fixedrange` property of the y-axis, which we set to `True`. This has the effect of locking the y-axis's range when zooming so that we can only change the x-axis range, which will make interactive exploration much easier. When we evaluate the preceding cell, we'll get the following figure:

Hourly pedestrian traffic for December 2019

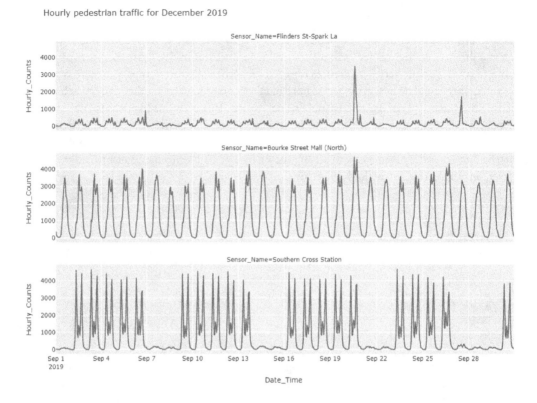

Figure 11.19 – A Plotly chart with three subplots, each containing a time series
plot of hourly pedestrian counts for a distinct sensor in September 2019

By plotting the hourly traffic of these three sensors in subfigures that share the same *x* and *y* axes, we're now able to readily compare pedestrian traffic patterns across them for September 2020. Comparing these three sensors, we can see that while they all have daily peaks and troughs, there is a noticeable variation:

- The Southern Cross Station sensor time series plot is quite striking, revealing a pronounced difference in weekday traffic versus weekend traffic, with the weekend daytime peaks being much lower than those on the weekend. This suggests that a large portion of pedestrian traffic comes from people commuting to and from work, which is consistent with the Southern Cross railway station being a major transport hub for Melbourne.

- The daily peaks for the Bourke Street Mall (North) sensor look similar in range to the previous sensor but with less pronounced variation between weekday and weekend traffic. This is likely associated with a large number of retail hubs along Bourke St mall, making this a popular weekend destination.

- Lastly, we can see that the `Flinders St-Spark La` sensor, which is located on the very edge of the central business district, has a much lower range of daily traffic counts, with discernible drops over the weekend. We can also see a large spike in pedestrian traffic on September 20, which can be explained by a Global Climate Strike that was held on this day, where the route of the protestors' march included a portion of Flinders St.

In the context of exploratory data analysis, a particularly convenient feature of the faceted plots that Plotly makes is that when using the zoom feature on a single subfigure, the range of the axes remains synchronized across all subfigures – though for this figure, we can only apply zoom to the x-axis due to the `fixedrange` property being enabled for the y-axis. This allows us to easily compare traffic through these sensors across narrower windows of time, without having to re-query our database or modify the configuration of our figure. You might want to use this to explore different time ranges and see if you can identify any other interesting patterns.

Our faceted time series plots have allowed us to identify some interesting, broad, and descriptive insights around daily pedestrian traffic changes. In the next section, we'll get a little more quantitative in our analysis and look at the statistical properties of the distribution of hourly pedestrian counts.

Visualizing the distribution of hourly pedestrian traffic

Time series plots are good at identifying broad seasonal and daily trends and are also useful for spotting anomalies. However, they are limited to plotting individual data points in time. If we want to draw some deeper insights into how pedestrian traffic varies across central Melbourne (and over time), we'll need to move away from visualizing individual data points to visualizing aggregated statistics across multiple data points. A frequently used plot for visualizing the distribution of a collection of values is a **box plot**. This plot provides a graphical representation of which value the dataset centers around, how spread out the values of the dataset are, and whether the dataset is skewed toward a particular end of the distribution. If you haven't worked with them before, you may find the following Wikipedia article on box plots a helpful reference as you work through this section: `https://en.wikipedia.org/wiki/Box_plot`.

Let's create a set of box plots that compare the distributions of pedestrian counts for a single sensor across multiple years. Since we already know that there is considerable variation across the hourly readings for each day, we'll look at the distributions of daily pedestrian counts so that we can focus on differences in the total number of pedestrians across a sensor per day. We'll look at traffic across the *Bourke Street Mall (North)* sensor, comparing the distributions for each year from 2019 to 2021.

To create these box plots, we'll need an extract of our `pedestrian_counts` table that contains daily count records for each day across the three years. To do this, we'll write a SQL query that groups the existing hourly records by both the year and the date components of the record's timestamp and then aggregates each year-day group by the sum of its hourly counts. To derive a new `Day` column for the grouping, we can simply cast it from a `TIMESTAMP` value to a `DATE` value:

```
%%sql bourke_daily_df <<
SELECT
    Year,
    Date_Time::DATE AS Date,
    sum(Hourly_Counts)::BIGINT AS Daily_Counts,
FROM pedestrian_counts
WHERE Sensor_Name = 'Bourke Street Mall (North)'
    AND Year IN (2019, 2020, 2021)
GROUP BY Year, Date
```

Let's inspect the resulting dataframe to double-check our data is in the right shape:

```
bourke_daily_df.head()
```

This gives us the following output:

	Year	Date	Daily_Counts
0	2019	2019-01-03	35470
1	2019	2019-01-04	22161
2	2019	2019-01-10	31016
3	2019	2019-01-13	27226
4	2019	2019-01-21	30637

Figure 11.20 – A Pandas dataframe showing a sample of five records of daily pedestrian counts for the Bourke Street Mall (North) sensor from 2019 to 2021

Success! This is in the right shape for us to make our box plots. We'll use Plotly's `px.box()` function, which, as you might imagine, allows us to make Plotly figures with box plots. We'll configure it to use the `Daily_Counts` column for our y-axis; to create separate box plots for each year, we'll use the `Year` column for our x-axis. We'll also set the points parameter to the `all` string, which instructs Plotly Express to also display the underlying data points alongside each box plot:

```
px.box(
    bourke_daily_df,
    x="Year",
```

```
      y="Daily_Counts",
      points="all",
      height=600,
      title="Distributions of daily traffic for a sensor",
)
```

When we evaluate this cell, the resulting figure object produces the following visualization:

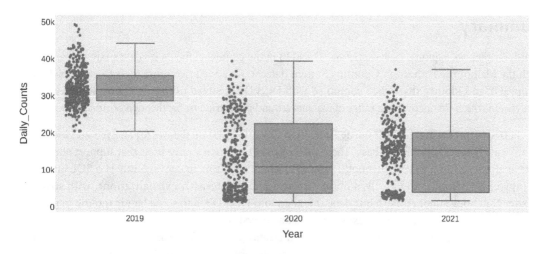

Distributions of daily traffic for a sensor

Figure 11.21 – A Plotly chart with three box plots comparing the distributions of daily
pedestrian counts from 2019 to 2021 for the Bourke Street Mall (North) sensor

These three box plots provide us with a rich set of information about the distribution of daily pedestrian counts across the three years. Hovering over each box plot informs us of the value of key summary statistics, including that they include the median daily count (the horizontal line inside the box), the first and third quartiles (the bottom and top of the boxes, enclosing 50% of the data points), and the lower and upper fence values (the horizontal lines bounding the whiskers extending from the box), which indicate sentinel data points, below or above which any data points are considered outliers. For the box plots of these three years, we can't see any outlier data points occurring outside the whiskers.

Unsurprisingly, the pre-COVID-19 distribution of daily counts for 2019 sits much higher than the next two years. We can also see that this year has a lower degree of spread than the other years, with the shorter box indicating a lower interquartile range – within which 50% of all the daily counts occur. 2020 and 2021 saw a greater spread of values, which is consistent with there being alternative periods of lockdown in both years. We can also see that 2020 has a slight skew toward lower values, and 2021 has a greater degree of skew toward higher numbers. This, as well as the slightly lower interquartile range for 2021, is consistent with 2021 having fewer days of lockdown than 2020.

As you can see, this dataset provides a wealth of possible insights into changes and variability in pedestrian traffic through Melbourne. We'll have to leave our exploratory analysis of the dataset here, but perhaps, while going through these analyses, you might have thought about different lines of exploration and questions that we could have pursued. Now that you have all the ingredients you need, we'll hand over the keyboard to you so that you can continue the analysis. This would also be a good time to look for opportunities to get some hands-on practice with other DuckDB SQL features that we covered in previous chapters, as well as other features from DuckDB's rich SQL API that we didn't have room to cover in this book.

Summary

In this chapter, we explored how we can use DuckDB in the context of hands-on data analysis. Working with the Melbourne Pedestrian Counting System dataset, we added two more tools to our toolchain – JupySQL and Plotly – that, when combined with DuckDB, enabled us to perform exploratory data analysis, in which we uncovered a range of insights around pedestrian traffic through central Melbourne.

We started by preparing the Melbourne Pedestrian Counting System dataset for analysis and loading it into a persistent DuckDB database. Then, we looked at two open source tools that support effective data analysis within Jupyter Notebooks: JupySQL, which allows us to conveniently run SQL queries in Jupyter Notebooks, and Plotly, a library for producing interactive visualizations, with strong Jupyter Notebook support. With our dataset loaded into DuckDB, and some handy tooling in place, we jumped into performing some exploratory data analysis of the Melbourne Pedestrian Counting System dataset. We set up a pattern of querying our cleaned dataset table using SQL JupySQL cells that produced pandas dataframes containing data extracts that we then fed into Plotly Express functions to produce our desired visualizations. Using this pattern, we looked at the total pedestrian counts across each year, noticing that we needed to account for the changing numbers of sensors to more accurately model and visualize this metric. Then, we compared the month-by-month traffic counts across 2019 to 2020 to dig deeper into the impact of COVID-19 lockdowns. After, we looked at time series plots of hourly counts, both for an individual sensor and across multiple sensors using a faceted plot. This enabled us to paint a much more detailed story of the impact of COVID-19 in 2020, as well as see how the rhythm of hourly pedestrian traffic varies throughout the day and can also differ between weekdays and weekends. Lastly, we performed a more quantitative analysis of pedestrian traffic changes over 2019 and 2021 by using box plots to compare their differing distributions of total daily traffic.

It's worth reflecting on how we interacted with DuckDB throughout this journey. When we prepared and loaded our dataset into a DuckDB table, we composed the necessary query using the DuckDB Python client's Relational API. When we moved on to querying this table for our data analysis, we switched to writing our SQL queries directly. JupySQL took care of dispatching these to our DuckDB database connection. There is no "right" way to arrive at this decision; we could have used either approach exclusively, or we could have even mixed them up. The best choice is the one that will make you and your team the most effective. Some factors to consider are that, as we discussed in *Chapter 8,* the composability of the Relational API via Python expressions often makes it a better choice for building out data applications using DuckDB. On the other hand, some more complex analytics queries may be easier to work with in SQL, or sometimes, this may also be necessary, given that the Python Relational API does not (at the time of writing) offer full support for the DuckDB SQL API.

Having worked through this chapter, you've now seen some tangible real-world examples of how DuckDB can be used to drive effective exploratory data analysis. The tools and workflows that we put into action are ones that you can use and incorporate into your data analysis toolkit.

In the next chapter, we'll conclude our DuckDB journey by offering a glimpse of the wider DuckDB ecosystem and looking at applications involving DuckDB clients, as well as some other tools that integrate nicely with DuckDB, including some that you might like to consider adding to your data analysis toolkit.

12

DuckDB – The Wider Pond

In this chapter, we'll turn our attention further afield from DuckDB itself to tools and services in the wider data ecosystem that integrate well with DuckDB. This will include open source packages and cloud services that enable more effective use of DuckDB for development and analysis, as well as more specialized libraries for building data applications, such as dashboards and other types of interactive data apps. We'll conclude this chapter – and our journey getting started with DuckDB – by covering a range of online resources that will be useful for diving deeper into DuckDB and engaging with the enthusiastic community that has formed around it.

In this chapter, we'll cover the following areas of the wider DuckDB and data ecosystem:

- Tools and services for working with DuckDB

- Libraries and interfaces that enable DuckDB integration

- Alternative DuckDB query interfaces

- DuckDB-powered data apps

- Keeping up to speed with the DuckDB ecosystem

By the end of this chapter, you'll be familiar with a range of different tools and services that can be used to augment and uplift DuckDB-oriented workflows, both for analysis and development. You'll also be equipped with a range of valuable resources to continue your adventures working with and leveraging DuckDB effectively.

Before we jump in, a comment on the scope of this chapter is in order. We have included a wide range of information in this survey of the extended DuckDB ecosystem, covering some rather distinct kinds of libraries and services that offer strong complementarity with DuckDB. While not every tool or service discussed may be immediately applicable to your current interests and projects, this overview is designed to familiarize you with the landscape, ensuring you are aware of these resources for when the need arises. The first and last sections are likely to offer more immediate value for supercharging your DuckDB workflows and adventures, while the sections in between may be relevant for more specific types of applications and workflows. You can use this chapter as a resource to come back to as you encounter relevant contexts in your analytical data endeavors.

Tools and services for working with DuckDB effectively

Throughout this book, we've largely worked with the DuckDB **command-line interface** (**CLI**) client when creating and running DuckDB SQL queries, except for our explorations of the Python and R clients. The DuckDB CLI is a convenient tool to have in your toolkit for quickly prototyping and experimenting with DuckDB queries, as well as for performing some tactical data wrangling. However, if you start working with DuckDB regularly, you'll likely encounter situations where a more fully featured development environment will enable you to work much more effectively. We've seen a couple of these already. In *Chapter 7* and *Chapter 8*, we saw how we can use the DuckDB Python client to work with DuckDB in a Jupyter Notebook, while in *Chapter 11*, we saw how the JupySQL library allows us to use DuckDB SQL directly via SQL-dedicated cells. Similarly, in *Chapter 9*, we saw how we can use the DuckDB R client to work with DuckDB databases in an R notebook environment.

There is a wide range of ways we can work with DuckDB beyond its CLI and notebook environments. In this section, we'll go on a brief tour of some tools in the data ecosystem that offer support for connecting to and working with DuckDB databases. This is not intended to be an exhaustive list; even if we tried to make it so, it would quickly become incomplete as this list is ever-growing! We'll cover the following types of tools:

- **SQL IDEs**: Tools for supporting SQL development workflows

- **Data explorers**: Tools that enable visual data exploration

- **Business intelligence** (**BI**) **tools**: More fully fledged libraries and applications that are designed to support and enable analysis of business metrics, including the development of reports, visualizations, and dashboards

- **Cloud services**: Fully managed services that provide environments for performing data analysis

Let's jump into this part of the DuckDB pond.

SQL IDEs

Just as there exists a range of **integrated development environments** (**IDEs**) for working with programming languages such as Python, C#, and JavaScript, there are specialized SQL IDEs (or SQL editors) that are designed to facilitate the development and running of SQL queries, as well as the management and administration of SQL databases. While their specific features vary, here are some typical features they offer that can improve developer efficiency and ergonomics when working with SQL **database management systems** (**DBMSs**):

- SQL-optimized code editing and syntax highlighting

- Query execution and result set management

- Database connection management

- Database catalog and records exploration

DBeaver (`https://dbeaver.io`) is a popular desktop SQL IDE that features strong DuckDB support out of the box. It has an open source community edition, as well as a commercial edition with enhanced functionality. See the DuckDB documentation for a guide on using DuckDB with DBeaver: `https://duckdb.org/docs/guides/sql_editors/dbeaver.html`.

For regular users of Visual Studio Code, the **SQLTools** extension is a popular SQL IDE, which (at the time of writing) has support for DuckDB via several community-supported drivers. You can find the SQLTools extension, as well as corresponding DuckDB drivers, via the **Visual Studio Marketplace**: `https://marketplace.visualstudio.com/vscode`.

Harlequin (`https://harlequin.sh`) is an open source SQL IDE that runs as a terminal application, like the DuckDB CLI, and has strong cross-platform support. Originally designed to work with DuckDB databases, Harlequin now supports a growing number of additional databases, including SQLite, Postgres, MySQL, MS SQL Server, and more. Harlequin gives you a range of powerful enhancements over the DuckDB CLI:

- A data catalog for navigating the contents of your database's tables, columns, and types, which can also display file trees for both local directories and S3-compatible remote cloud-object storage
- An SQL query editor that supports multiple query tabs and offers autocompletion, loading and saving queries, and query auto-formatting
- A results viewer for inspecting large query sets with an interactive tabular view
- The ability to export query results to CSV, Parquet, and JSON
- A query history viewer that you can browse and reuse in new query-editing sessions

These features make Harlequin a powerful drop-in replacement for the DuckDB CLI, giving you all the capabilities you'd expect from an SQL IDE, all from within your terminal:

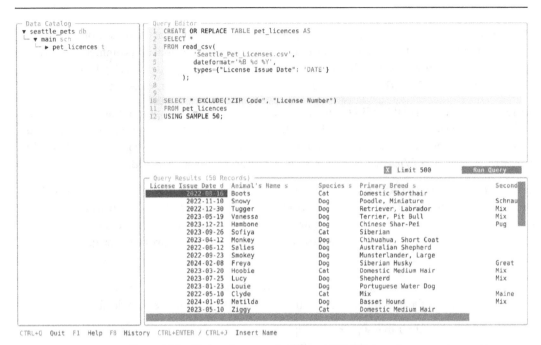

Figure 12.1 – Harlequin, a terminal-based SQL IDE, in action

Data explorers

Sometimes, you may want to rapidly explore a dataset to understand its contents. This could be when you're working with a new dataset that you're unfamiliar with, or perhaps it could be when you want to verify that a generated dataset has the structure you expect. In these situations, where the focus is on data exploration, rather than on developing repeatable analyses or data transformation pipelines, having to write code – whether it be SQL, Python, or R – to explore what's in your dataset can introduce unnecessary friction.

An alternative to writing code for dataset exploration is to use a visual no-code tool that is specifically designed to enable rapid data exploration. **Tad** (https://www.tadviewer.com) is one such data exploration tool with DuckDB support. Tad is a desktop application for viewing, exploring, and analyzing data, with strong support for creating pivot tables. It supports reading CSV and Parquet files, as well as DuckDB and SQLite databases. Not only does Tad support reading and exploring the contents of DuckDB databases, but it also uses DuckDB as its internal analytical engine, enabling it to create pivot tables and run queries efficiently in response to user interactions.

Another visual data exploration tool is **Visidata** (`https://www.visidata.org`). Rather than being a desktop tool, Visidata is a terminal-based application that enables you to quickly open, explore, summarize, and analyze datasets. It supports a wide range of input data formats, supporting any file format that pandas can read. This enables you to use Visidata to quickly open, explore, and even transform CSV, Parquet, JSON, and other file formats in-memory. Where DuckDB comes into the picture is that Visidata also comes with a companion tool called vdsql, which allows you to connect Visidata to a range of SQL databases, including DuckDB. This means that you can use Visidata to directly connect to a DuckDB database, and even explore and wrangle its contents using Visidata via operations that are exposed through keyboard commands. If you need to reuse the queries that are performed during your exploration, you can also extract the resulting SQL queries that are generated through the course of your interactive visual analysis. vdsql's multi-database support – allowing you to connect to other databases, including SQLite, PostgreSQL, MySQL, Clickhouse, Google BigQuery, and Snowflake – is enabled via the Ibis Python package. We'll cover this later in this chapter, in the *Alternative DuckDB query interfaces* section.

BI tools

BI tools are applications that support the collection, processing, and analysis of data from various sources to provide insights and support decision-making around the strategic and operational performance of an organization. They enable self-service interactive data analysis, as well as the development of reuseable dashboards and reports. DuckDB's capabilities as a performant **online analytical processing (OLAP)** database make it well-suited to being a data source for BI tools. As we'll see, some BI tools have also started making use of DuckDB for the underlying engine that's used for performing necessary data transformations and running interactive queries.

Two of the most popular BI tools out there are **Tableau** (`https://www.tableau.com`) and **Power BI** (`https://powerbi.microsoft.com`). They offer broadly comparable features, with each having different strengths. Tableau is known for having strong visualization capabilities and supporting more advanced charting configurations, whereas Power BI benefits from good integrations within the Microsoft ecosystem and has strong data modeling and data pipelining features. At the time of writing, these tools don't have official DuckDB integrations yet, but we expect this to change soon. In the meantime, Tableau can be connected to DuckDB via a community-provided open source connector, and Power BI can be connected via DuckDB's ODBC client, though you may run into some rough edges here and there. Consult DuckDB's documentation for the most up-to-date advice on how best to connect BI tools such as Tableau and Power BI to DuckDB.

Metabase (`https://www.metabase.com`) and **Apache Superset** (`https://superset.apache.org`) are both popular open source data visualization and BI tools with official support for DuckDB as a data source. Metabase has a user-friendly interface that is designed to enable self-service analytics for a wider range of personas that have varying levels of technical expertise. Apache Superset, on the other hand, is geared a little more toward data developers, such as data analysts, data scientists, and data engineers, offering an integrated SQL IDE for developing queries that power your dashboards. It has a slightly steeper learning curve but provides a richer set of data visualization features and more customizability, making it a good option for building more bespoke analytical dashboards and data apps.

Rill (`https://www.rilldata.com`) and **Evidence** (`https://evidence.dev`) are two more open source BI tools, both of which have good DuckDB data source integration. They also both make use of DuckDB internally for data transformations and running queries. They are notable for both having an emphasis on BI-as-code, with queries and dashboard configuration being defined as code, enabling the application of version control and testing automation practices. Rill places a particular emphasis on providing powerful data exploration features through a SQL IDE and tabular data viewers enriched with visual summary information. The results of your data exploration can automatically be converted into Rill's code-based configuration, which allows developers to go from exploring to deploying dashboards through one tool.

Like Rill, Evidence also emphasizes developer ergonomics but takes a more code-first approach to dashboard and report development, with their contents being specified by a lightweight syntax that includes Markdown, SQL, and visual charting components. Its programmability extends to supporting the dynamic generation of page contents from your data. Evidence is designed to support building a full spectrum of data apps, from static reports to interactive dashboards and custom decision-support tools. The open source Evidence project has strong support for building custom-templated reports that are generated from enriched data and can be published as static web pages. Recently, the project has added support for building data apps that support interactive querying and filtering, with dynamic query execution being driven by a DuckDB process running completely in the browser using the DuckDB-Wasm client. We'll discuss this pattern of running DuckDB in the browser to power data apps later in this chapter, in the *DuckDB-powered data apps* section.

DuckDB in the cloud

Throughout this book, we have been working with DuckDB running in a local environment. Another strategy that's open to data practitioners in need of interactive data analytics environments is to use an appropriate cloud-based service. Given DuckDB's high degrees of portability and simplicity to install, the number of places you can run DuckDB in the cloud is virtually limitless, and there is a large and growing number of data-oriented cloud service providers adding DuckDB as an integration feature into their products. Here, we will focus on a hand-picked selection of cloud services that provide fully-featured computational environments that are designed for working with data, and which DuckDB complements well. In addition to providing cloud-based computation environments across Python, R, and SQL, they all offer notebook-like interfaces for interactive data analysis and development. They are also all services you can dive into readily right now – provided, of course, you have an internet connection.

MotherDuck (`https://motherduck.com`) is a cloud-hosted analytics platform that's purpose-built around DuckDB, offering hosted DuckDB databases alongside a broad suite of features. As we have seen in multiple places in our journey, DuckDB databases can be backed by a persistent file-based database, which enables the reuse of databases over time and by multiple consumers. Teams making heavy use of persistent DuckDB databases may find that the management of these resources introduces complexity they would prefer not to manage themselves. MotherDuck's platform aims to streamline these concerns for practitioners and teams by managing the storage, performance, organization, and availability of a collection of DuckDB databases. The platform offers a centralized data catalog that combines tables from across multiple databases, as well as other features of data warehousing platforms, such as integration with data ingestion, orchestration, and **change data capture** (CDC) tools. MotherDuck also offers an integrated SQL-based notebook environment, which is designed to enable rapid data exploration and analysis using DuckDB, coupled with a collection of automated visualizations and summary statistics. A core design principle of Motherduck's platform is enabling hybrid workflows across both local and remote DuckDB databases. Users can query MotherDuck-hosted databases from DuckDB running on their local machine and can also run hybrid queries across both local and MotherDuck DuckDB databases. MotherDuck offers an enhanced query planner, which enables hybrid execution across both local and remote sources of compute, distributing the load dynamically at query time to better leverage available resources and reduce query times. This range of versatile features that build on and augment DuckDB's capabilities makes MotherDuck a compelling service both for individual practitioners and data teams looking to make effective use of DuckDB for both their ad hoc and operational data workflows, as well as for powering DuckDB-powered data products.

Google Colab (`https://colab.google`) is a hosted Jupyter Notebook service that allows users to work with both Python and R-based notebooks running in a managed environment that's pre-installed with a range of data analytics and data science libraries, all of which you can extend. This makes it a good service for working effectively with DuckDB in a Jupyter Notebook, just as we did with a local Jupyter Notebook in *Chapters 7*, *8*, and *11*. Google Colab also gives you the ability to use accelerated compute instances to run your kernel, which is necessary for a range of data science and **machine learning** (ML) applications. If you have a Google account, it is easy to get started at `https://colab.research.google.com`, or by choosing *Google Colaboratory* when you create a new document in Google Drive.

Kaggle Notebooks (`https://www.kaggle.com/docs/notebooks`) are cloud-hosted computational notebook environments that are designed to enable collaborative analysis and the development of reproducible artifacts. This is a core feature of the Kaggle platform and community, which brings together ML practitioners to learn, practice, and share knowledge. While Kaggle is best known for being an ML-oriented community, it so happens that data analysis and data preparation are both crucial parts of the ML development life cycle, and many high-quality notebooks contributed by the community focus on data analysis and strategies for preparing datasets for use with ML models. Whether you're applying ML techniques or not, Kaggle Notebooks are great environments for using DuckDB for interactive data analysis, along with a range of other complementary data tools. Like Google Colab, Kaggle Notebooks offer support for both Python and R notebook kernels and allow notebooks backed by accelerated compute-enabled environments to be run.

Binder (`https://mybinder.org`) is another Jupyter Notebook-related cloud service, but what it does is a little different to both Google Colab and Kaggle Notebooks. Rather than giving you a pre-defined environment with a common set of packages already installed, Binder takes the complete opposite approach by allowing you to define a custom environment precisely as required for your data analyses or data app hosting needs. To use Binder, you simply need to point it at a GitHub repository that contains one or more Jupyter Notebooks, and Binder will take care of spinning up an appropriate cloud computing environment with a full JupyterLab (or classic Jupyter Notebook) session ready for you to work with. You can also provide configuration within your repository with instructions to Binder for setting up any required dependencies for your environment, such as which packages should be installed, datasets that should be downloaded, and which kernel should be used to run a notebook – allowing you to use a wide range of runtimes beyond the standard Python and R options. Binder is a free-to-use service and only requires you to provide a URL to a publicly accessible GitHub repository to launch an environment. It is important to note that any environments you spin up will only be temporary and will spin back down after a period, meaning they are not suitable for long-running workloads. However, they are perfect for short-lived customized data science and analytics environments, in which you can make use of DuckDB, along with virtually any other open source data library you can run via a Jupyter Notebook.

Observable (`https://observablehq.com`) is another hosted data analytics environment that offers a fully featured notebook interface. Observable notebooks allow you to define code cells containing either JavaScript or SQL code for defining data queries, transformations, and interactive visualizations. Observable comes with strong integration for **Observable Plot** (`https://observablehq.com/plot`), an open source data visualization library that's designed to accelerate the development of rich data visualizations, especially in the context of exploratory data analysis, where rapid iteration is crucial. Observable offers a suite of collaboration features, including notebook sharing, inline commenting on notebooks, and the ability for multiple editors to collaborate on a notebook simultaneously. Observable notebooks can also be configured to consume from a range of data sources, such as SQL databases, cloud file storage, and cloud object storage. Of particular interest is Observable's DuckDB database client, which is based on DuckDB's WebAssembly client, and runs entirely within the browser. This allows you to query attached files using DuckDB SQL from SQL notebook cells to drive responsive exploratory data analysis and visualization.

Having looked at a range of DuckDB-friendly tools and services for SQL query development, database management, data exploration, and BI analysis, we'll turn to look at packages and standards that facilitate integration and improved modularity between DuckDB and other data-producing and data-consuming tools in the data ecosystem.

DuckDB integration

In this section, we'll look at some different ways you can enable your data-oriented applications to integrate with DuckDB as a data source. DuckDB can be used via a range of various client APIs across different languages. At the time of writing, the DuckDB documentation lists 14 officially supported client APIs across different programming languages and runtimes, and seven third-party client APIs (`https://duckdb.org/docs/api/overview`). When an application or library needs to connect to a collection of databases, which may need to grow over time, integration via database-specific APIs will add considerable development and maintenance overheads each time a new database is added. In such contexts, you'll likely want to consider adopting a standardized interface that will allow you to target a single API that provides support for integrating across multiple data backends. In this section, we'll outline some standards and tools that DuckDB integrates with to help streamline the development of applications involving multiple data backends.

ODBC and JDBC

Open Database Connectivity (**ODBC**) and **Java Database Connectivity** (**JDBC**) are standardized interfaces for connecting and interacting with databases. They both provide a consistent way to access and interact with data stored in various database management systems, and they are both frequently used to support database integration for BI tools, reporting software, and data analysis applications. ODBC is designed to be used with a range of language runtimes, whereas to use JDBC, your application must be using Java or another language based on **Java Virtual Machine** (**JVM**). They both provide varying degrees of cross-platform support. For an application to connect to a database using either ODBC or JDBC, a target database driver for the database needs to have been implemented. DuckDB provides official clients for both JDBC (`https://duckdb.org/docs/api/java`) and ODBC (`https://duckdb.org/docs/api/odbc/overview`).

ODBC and JDBC are important workhorses in the data ecosystem, enabling interoperability between a wide range of analytical data applications and many data processing engines. While they address the challenge of integration, they come with notable performance trade-offs when used for analytical workloads. In particular, they are geared toward providing row-oriented result sets, in contrast to columnar data structures, which offer greater performance for analytical workloads. When connecting to an analytical database or query engine via ODBC or JDBC, this means that you will likely pay significant overheads when serializing query results into a row-oriented data structure, only be to deserialized back into an analytical-data-friendly structure in the client application.

ADBC

Arrow Database Connectivity (**ADBC**: `https://arrow.apache.org/docs/format/ADBC.html`) is a standardized database interface that is built around **Apache Arrow**. Apache Arrow is a columnar in-memory data format that's designed for analytical data applications. Its column-oriented data structure enables query engines to leverage fast vectorized computation, which requires that column records be located contiguously in-memory. The ADBC standard enables further analytical querying performance enhancements, such as projection pushdown, where columns that are not needed for a given query can be skipped at read time, resulting in reduced data movement. Apache Arrow is becoming increasingly ubiquitous across the data ecosystem, which, coupled with its standardized format, means that the same in-memory data structure can be used across clients and over the network, without you having to pay serialization and deserialization costs, and, sometimes, without having to copy any data.

Being built around Apache Arrow, ADBC can address the inefficiencies of ODBC and JDBC. When using an ADBC driver to connect to a database that has native Arrow support, the driver can simply pass through the same bytes returned from the database, without having to perform any costly serialization and deserialization of results. When the database and querying process are located on the same machine, ADBC drivers can even support zero-copy operations as they only need to pass a pointer to the memory location of the results. For connectivity with databases that don't support Arrow output, ADBC drivers can apply the necessary conversion into Arrow tables.

DuckDB provides official support for ADBC via its ADBC client (`https://duckdb.org/docs/api/adbc`). At the time of writing, the ADBC standard is relatively new, and it does not have the ubiquitous support (in terms of database connectors and consuming applications) that ODBC and JDBC do yet. With the increasing adoption of Apache Arrow across the data ecosystem, we believe it is only a matter of time before ADBC becomes a go-to standard for analytical data interconnectivity.

Substrait

Substrait (`https://substrait.io`) is an open source format for specifying operations on structured data that can be run across multiple languages and systems. It aspires to be a universal interchange format for defining query plans that can be produced by any Substrait-compatible query interface and consumed and run on any Substrait-compatible database engine. While Substrait is still an emerging standard in the data ecosystem, as more systems start to add Substrait query plan production and consumption, the benefits of being able to reuse the same query across multiple execution engines – and generally not having to solve the problem of translating queries for point-to-point integrations – will become increasingly tangible. The Substrait project consists of a formal specification, a human-readable text representation, and a compact cross-language binary representation.

DuckDB can both produce and consume Substrait query plans through an official DuckDB extension (`https://duckdb.org/docs/extensions/substrait.html`). This means that DuckDB can be used to consume and execute queries created in another tool that emits Substrait query plans, such as Ibis or Apache Arrow DataFusion, and conversely, Substrait query plans generated by DuckDB can be run on engines that can consume Substrait. This includes Apache Arrow Data Fusion and Apache Acero, with many more to come (we hope). Note that, at the time of writing, DuckDB's Substrait integration is experimental, and developer support is only available on a limited basis.

SQLAlchemy

SQLAlchemy (`https://www.sqlalchemy.org`) is an open source library for working with SQL and SQL databases in Python. SQLAlchemy is made up of two high-level components:

- **SQLAlchemy Core**: A foundational set of components that allow you to manage collections of database connections, database schemas, and SQL expressions, all from within Python.

- **SQLAlchemy Object Relational Mapper** (**ORM**): SQLAlchemy's ORM builds on the foundations of SQLAlchemy Core to provide a mapping from Python classes to database tables, and a resulting mapping from instances of those classes to corresponding table rows. This allows you to work with database records via Python objects that model their structure.

SQLAlchemy's ORM is particularly well suited to supporting stateful changes to databases via Python application logic. This is more commonly seen in **online transaction processing** (**OLTP**) workloads, such as managing users' state within web applications. It is less well suited, however, to OLAP workloads, where it is common for queries to return very large numbers of rows. Retrieving DuckDB query results through SQLAlchemy's ORM would require that each record within a result set be converted into a new Python object, resulting in significant memory and processing speed inefficiencies. So, while it is possible to use SQLAlchemy's ORM with DuckDB, the real value is to be had from SQLAlchemy's Core components, in particular its database connection management features, and potentially also its SQL expression interface.

DuckDB's integration with SQLAlchemy is enabled through the community-supported open source library **DuckDB Engine** (`https://github.com/Mause/duckdb_engine`). This provides a SQLAlchemy dialect for DuckDB, which allows you to create SQLAlchemy engines that connect to DuckDB. Each SQLAlchemy Engine is composed of a database-specific dialect, and a pool of DB-API connection objects that support the maintenance of long-running connection objects in memory. As we discussed in *Chapter 8*, Python's DB-API is a specification for a standardized interface for communicating between Python applications and databases. While the DB-API specification improves consistency across Python database integration, there is typically some variation across DB-API implementations for different Python libraries, meaning that a small amount of integration logic is often required for different databases. Using SQLAlchemy engines to connect to target databases provides enhanced integration consistency across different databases, freeing up developers from having to manage database-specific integration needs.

These integration features make SQLAlchemy a useful target for connecting DuckDB to both custom Python applications and data tooling libraries. For example, SQLAlchemy (via DuckDB Engine) provides integration for the JupySQL (which we covered in *Chapter 11*), Superset (which we discussed earlier in this section), and Ibis (which we will discuss in the next section) libraries in the Python data ecosystem.

In the next section, we'll look at a different form of modularity to be found within the data ecosystem involving DuckDB. More specifically, we'll look at alternative code-based querying interfaces that can target DuckDB. These libraries enable you to write reusable queries that can be executed across multiple data backends, including DuckDB.

Alternative DuckDB query interfaces

While DuckDB's primary interface is DuckDB SQL, this isn't the only way you can compose DuckDB queries. In *Chapter 8*, we saw how the DuckDB Python client's Relational API allows you to use a dataframe-like API to create relation objects, which can be composed through method chaining to effectively define complex DuckDB queries. In *Chapter 9*, we also saw how, when working in R, DuckDB supports executing queries composed with dplyr, through the composition of piped function calls. In this section, we'll survey some notable query interfaces that can be used as alternatives to SQL for querying DuckDB.

One motivation for adopting an alternative query interface is that many data practitioners, especially those who work in Python and R, may prefer not to use SQL to define analytical queries. Furthermore, these query APIs are designed with analytical query patterns in mind and can often express certain types of analytical queries much more concisely and ergonomically than they would be in SQL, resulting in simpler and more maintainable query code.

Another benefit of adopting an alternative query interface to SQL is portability. Queries defined in DuckDB SQL can only be executed by DuckDB. Most of the query interfaces we'll look at support multiple backends as their query interfaces. This allows data practitioners and teams to adopt a single preferred query interface that is portable across a range of query engines beyond DuckDB. This improves the reusability and maintainability of query logic as data teams can maintain a single collection of queries rather than having to translate and maintain different sets of queries for different execution engines. It also provides data teams and practitioners with more flexibility around where they run their queries. For example, when a compute-intensive set of data transformations needs to be scaled up to be run over larger datasets than can be handled by a single machine, the same queries can be used to target a scalable distributed query execution engine. Let's go through some notable query interfaces for working with analytical data that have DuckDB integration.

dplyr

dplyr is an R library that provides a dataframe-oriented interface for composing analytical queries and data transformations through a composable grammar of data manipulation. In addition to providing an ergonomic and expressive data manipulation interface, dplyr is designed such that the description of data manipulation is decoupled from its actual execution, meaning that the same dplyr code can be used to query a range of data backends, beyond just R dataframes. This includes a range of SQL databases, including DuckDB. In *Chapter 9*, we saw how, when working in R, we can use dplyr as an alternative interface for querying DuckDB. This is made possible through two components:

- The dbplyr package, which allows you to convert dplyr code into database-specific SQL.

- The DuckDB R client, which provides support for R's **Database Interface** (**DBI**). As we discussed in *Chapter 9*, the DBI provides a standardized interface for database-specific drivers, allowing you to connect to and query a range of different databases from within R. The R DuckDB client provides the necessary configuration that enables DuckDB to be used as a dbplyr backend.

Using DuckDB as the dplyr backend (via dbplyr) allows R users to compose analytical queries effectively using dplyr's elegant syntax, while also being able to take advantage of the many compelling features of DuckDB. Compared to the default dataframe backend, using DuckDB to execute dplyr queries enables dramatically faster query execution and the crunching of much larger datasets. dplyr's support across a wide range of databases means that you can use the same dplyr query code against other SQL databases beyond DuckDB. This portability of queries can be particularly convenient when you're going from a local prototype to a production SQL database and can also make it easier for a data team to make the strategic decision to change which execution engine they use for operational workloads.

A large contributor to dplyr's popularity among R users is its thoughtfully designed interface, which has been designed with the needs of analytical data transformations in mind. Its composable grammar of graphics provides users with an intuitive and consistent interface that can concisely express analytical operations, including queries that are infamous for requiring complex and impenetrable SQL queries. Thanks to dplyr's modular design, and DuckDB's support for dbplyr, R users can take advantage of the performance-enhancing benefits of DuckDB, while still being able to use their preferred query interface.

duckplyr

Using dplyr to query DuckDB databases via the dbplyr backend is a viable way to leverage the combined benefits of dplyr and DuckDB. However, it does come with some trade-offs. First, dplyr queries must be translated into DuckDB SQL before they can be executed against DuckDB. This introduces a level of indirection between the user-facing dplyr interface and DuckDB's internal execution engine. When errors are encountered at the execution layer, DuckDB can only surface diagnostic feedback at the SQL level, which can make it challenging for the user to establish which parts of their dplyr code errors originate from, especially for users who predominantly work with dplyr rather than SQL. Additionally, the translation of R operators, identifier names, constants, and table names cannot always be done reliably. Users may need to identify and address such conversion issues through close inspection of the translated SQL.

duckplyr is an R package that is designed to be a drop-in replacement for dplyr, allowing users to run dplyr queries against DuckDB databases. Rather than translating dplyr expressions into DuckDB SQL, duckplyr uses a DuckDB-native interface. Bypassing the SQL translation step entirely and targeting DuckDB's internal API provides much cleaner integration with DuckDB, eliminating translation errors as well as offering superior diagnostic feedback around query errors. Native execution also results in substantial performance improvements compared with SQL translation. To achieve duckplyr's goal of being a reliable drop-in replacement for dplyr, only a selected subset of dplyr's functions, as well as R functions and data types, are currently implemented. On encountering an unsupported feature, duckplyr falls back to using dplyr, ensuring that your query will still run, albeit with reduced performance.

duckplyr is readily available for installation via **Comprehensive R Archive Network** (**CRAN**), R's primary repository for R packages. Installing it is as simple as running the following command:

```
install.packages("duckplyr")
```

duckplyr is under active development, with further native coverage over dplyr and R language features on the roadmap, but it is ready for use now. If you find the combination of dplyr and DuckDB compelling, we recommend exploring duckplyr.

Ibis

Ibis is a Python library for defining data transformations using a dataframe-like API that is well suited to data analytics and data science workflows. It draws inspiration in its design from both pandas and dplyr; users of both tools will find familiar elements. Where the query interface for the pandas and Polars dataframe libraries are coupled to their implementation, the Ibis API is not coupled to a specific execution engine. Ibis does not perform data transformations itself; instead, it translates Python expressions into SQL queries across a range of SQL dialects corresponding to each supported backend. In this way, it is analogous to using dplyr with the dbplyr backend, which allows you to produce backend-specific SQL from dplyr queries. You can think of Ibis as filling a similar position in the Python ecosystem.

Ibis gives you the expressiveness of a dataframe library – allowing you to define complex analytical queries much more effectively than in SQL – while also giving you the performance benefits of modern SQL engines. Ibis supports both in-process backends (DuckDB, pandas, and Polars) as well as distributed data processing engines. This support for both kinds of execution backends allows you to prototype and develop queries over smaller datasets on a local machine and then jump to a much more scalable execution engine when needed. DuckDB is the default backend for Ibis, which is a great combination for a local development workflow, allowing you to take advantage of DuckDB's ability to scale many local workloads beyond what dataframe libraries can handle. When you need to go from prototype to production, or when your data volume grows dramatically, beyond what DuckDB can support, you can take the same Ibis queries, with minimal or often no changes, and scale them out by targeting a distributed processing engine such as Apache Spark, Snowflake, or BigQuery.

In addition to enabling smoother prototyping workflows, this portability can also help minimize change management risk by allowing organizations to replace data processing engines without requiring a resource-intensive migration process for query code. This also means that data practitioners won't have to go through the process of learning a new query interface when they're changing from one execution backend to another, reducing the impact of these changes on productivity.

Getting hands-on with Ibis

Let's take a very brief look at what using Ibis to compose queries for execution in DuckDB looks like. We'll use the NYC Dog Licensing dataset, which we explored with R and DuckDB in *Chapter 9*. For this example, you'll need a Jupyter Notebook-compatible IDE that's been set up with the same Python environment we used for *Chapters 7*, *8*, and *11*.

Ibis should already be installed in your environment due to you following the steps in the *Technical requirements* section of *Chapter 7*. So, after opening chapter_12_ibis.ipynb, you'll be ready to import Ibis and create a connection to a new in-memory DuckDB database:

```
import ibis
conn = ibis.duckdb.connect()
```

Ibis connection objects are instances of Ibis Backend classes that correspond to a given backend, and which expose backend-specific functionality. You can find documentation around Ibis' DuckDB Backend class here: https://ibis-project.org/backends/duckdb. We'll use the connection's read_csv() method – a wrapper around DuckDB's read_csv function – to read the dataset's CSV file located in the chapter_09 directory:

```
licenses = conn.read_csv(
    "../chapter_09/NYC_Dog_Licensing_Dataset.csv",
    ignore_errors=True
)
```

Our licenses variable now contains an Ibis Table expression representing the loaded contents of the CSV file. Much like DuckDBPyRelation from the DuckDB Python client, Ibis Table expressions are lazily-evaluated representations of queries, so we have yet to read any results from the file. They differ from DuckDB's relations, which are part of DuckDB's internal API, in that when they're executed, Table expressions are translated into DuckDB SQL before being submitted to the database for execution, much like when using dplyr and dbplyr to query a DuckDB database.

The Ibis API encourages the use of chaining Table expressions to compose queries made up of multiple operations, much like we saw with the Relational API in DuckDB's Python client in *Chapter 8*, and with dplyr in *Chapter 9* – though dplyr exposes chaining via piped function calls rather than method calls. To readily build Table expressions that define operations over columns, Ibis supports the convenient creation of column expressions using its underscore (_) API:

```
from ibis import _
```

We now have the ingredients we need to recreate the data preparation steps we performed with dplyr in *Chapter 9*. This involved filtering out records with missing values and filtering down the dataset to a single record for each dog in the dataset by removing repeat license records for each pet. We'll do this by building up a corresponding `Table` expression, starting with the `licenses` expression:

```
distinct_cols = [
    "AnimalName", "AnimalGender", "AnimalBirthYear",
    "BreedName", "ZipCode"
]
unique_dogs = licenses[distinct_cols].filter(
    _.AnimalName.notin(
        ["UNKNOWN", "NAME NOT PROVIDED", "NAME", "NONE"]
    )
).filter(
    _.BreedName.notin(["Unknown", "Not Provided"])
).distinct()
```

In this code, we've created an Ibis `Table` expression and assigned it to the `unique_dogs` variable, which contains all the operations involved in our preparation steps:

1. We started by dropping the columns we don't need with the `licenses[distinct_cols]` expression, which selects only columns in the `distinct_cols` list.

2. Then, we used the `filter()` method to remove records with missing or junk values in the `AnimalName` column. The `_` identifier provides a convenient way to define a column expression that captures the logic we need for this.

3. Another chained `filter()` call allows us to perform a similar cleaning step for the `BreedName` column.

4. Finally, we apply a `distinct()` method call, which will remove duplicate rows sharing the same values across all columns (which have already been reduced to only those in the `distinct_cols` list).

We now have a `Table` expression in the `unique_dogs` variable that represents all the operations in our data preparation step.

Let's use our `unique_dogs` expression to compose a query that counts the number of dogs born in each year, sorting our results in descending order of birth year. If we were composing a SQL query, we would use a GROUP BY clause over `AnimalBirthYear`, combined with a `count` aggregate function, adding an ORDER BY `AnimalBirthYear` DESC clause. In terms of our Ibis expression, we can achieve this by chaining the `group_by()`, `aggregate()`, and `order_by()` methods:

```
dogs_by_year = unique_dogs.group_by(
    "AnimalBirthYear"
).aggregate(
```

```
    Count=_.count()
).order_by(
    ibis.desc("AnimalBirthYear")
)
```

We now have a `Table` expression in the `dogs_by_year` variable that contains our desired set of operations. By default, evaluating `Table` expressions in your Python interpreter will just provide information about the expression, rather than triggering execution. We can change this by enabling Ibis' interactive mode, which will make Ibis partially execute queries, giving us a preview of the results. Now, when we call `head()` on our expression, we'll see the first set of results from the partial execution of the resulting DuckDB query:

```
ibis.options.interactive = True
dogs_by_year.head(10)
```

Evaluating this produces the following output:

AnimalBirthYear	Count
int64	int64
2023	3799
2022	10943
2021	16981
2020	23241
2019	19000
2018	19529
2017	20565
2016	20249
2015	19063
2014	16984

Figure 12.2 – The first 10 rows from our aggregate Ibis Table expression

Finally, we will call the `to_pandas()` method that `Table` expression objects provide, which will fully materialize the query results for this expression, returning a pandas dataframe containing all the results for this query:

```
dogs_by_year_df = dogs_by_year.to_pandas()
```

We will end this brief exploration of the Ibis API here as we have further alternative query interfaces for DuckDB to cover. Hopefully, this has given you a sense of how Ibis offers a compelling dataframe-like API for interacting with not only DuckDB but other SQL-database backends. If you want to dive deeper into Ibis, the project's home page provides comprehensive documentation and a range of guides for hands-on usage: `https://ibis-project.org`.

Fugue

Fugue is similar to Ibis in that it is a Python library that aims to provide an effective interface for defining data transformations that can be executed on a range of backends, including both in-process and distributed data processing engines. While Ibis provides a single dataframe-like interface for producing SQL queries, Fugue gives you more flexibility in how you define your data transformations by offering two distinct APIs:

- **Fugue API**: An interface that allows you to run business logic defined using Python and pandas code against scalable distributed processing engines, including Apache Spark, Dask, and Ray.

- **Fugue SQL**: A SQL dialect that is built on top of the Fugue API, allowing users who are more comfortable with SQL to define data transformations that can be run on the same distributed processing engines. Fugue SQL extends traditional SQL with affordances for working with distributed workloads, such as the PERSIST and BROADCAST statements. It also provides enhancements that enable composable end-to-end data processing pipelines to be defined using a more concise syntax than traditional SQL, where complex query chaining can become difficult to manage.

These two APIs can be combined, with Fugue SQL being able to both query dataframes and also invoke Python and pandas code. Fugue also provides a %%fsql cell magic (like JupySQL's %%sql magic, which we saw in *Chapter 11*), allowing you to run Fugue SQL directly in Jupyter Notebooks. All these features of Fugue's APIs allow it to function as a data pipeline orchestrator by defining end-to-end data transformations across SQL and Python that can be run on both local machines as well as on Spark, Dask, and Ray clusters. This gives you some of the same portability benefits we saw with Ibis, avoiding the need for rewriting code when moving from local prototyping to a distributed engine, or when needing to switch between Spark, Dask, and Ray.

Fugue supports DuckDB as one of its execution engines for running local workloads, which will often be much more performant than using its native execution engine, which uses pandas. This means that you can prototype queries using the DuckDB execution engine, and then switch to running them against an execution engine for running on a cluster while using the same Fugue queries. A particularly useful workflow is sampling a large dataset from a cluster as a dataframe, and then working with that locally using the DuckDB execution engine, again, while using the same querying interface.

Having looked at the features and benefits of adopting alternative query interface tools for working with DuckDB, let's summarize the core motivations for you perhaps wanting to adopt one yourself:

- Writing queries once and not having to rewrite them again for use with other query engines, such as when needing to use different execution engines across local, testing, and production environments, or when migrating from one data execution engine to another

- You can leverage the benefits of non-SQL querying interfaces that provide better abstractions for analytical query patterns while still retaining all the performance benefits that DuckDB provides

The DuckDB Spark API

In *Chapter 8*, we looked at two different APIs that DuckDB's Python client supports: the DB-API and the Relational API. These offer different ways to interact with and issue queries to a DuckDB database. There is another API that the DuckDB Python client offers that we have not discussed yet – the DuckDB Spark API. This is an implementation of the PySpark API (the Python API for Apache Spark) that allows you to interact with DuckDB using PySpark queries. This is made possible by the DuckDB Python client through its Relational API, which translates PySpark queries into DuckDB internal query plans, which are then executed against the DuckDB query engine. This allows you to run the same PySpark code against either Apache Spark clusters or DuckDB running on a single compute node.

This increased flexibility around the types of environments that PySpark queries can be executed against opens a range of additional use cases for data teams who make heavy use of Apache Spark. There are three broad types of benefits of using DuckDB as a query engine for running PySpark queries that are worth calling out:

- Apache Spark is specifically designed to perform well over large datasets that cannot be processed in a reasonable time by a single machine. For smaller workloads, the performance benefits of using a distributed execution engine such as Spark are often negated by the overheads that are involved with serializing data partitions and transferring them to nodes in the cluster over the network. DuckDB's impressive performance means that for many small-to-medium datasets, DuckDB running on a single node will offer greater efficiency and speed compared to running the same queries against a Spark cluster. DuckDB's PySpark-compatible API means that data teams can leverage the most appropriate type of execution engine for their workload without needing to translate into and maintain a second collection of queries written against different APIs.

- DuckDB's PySpark-compatible API also allows data practitioners to rapidly prototype queries using a local DuckDB database before moving them to a Spark cluster, where the development feedback cycle will be slower.

- The ability to run PySpark queries effectively on a single machine expands the availability of quality assurance techniques for identifying issues during the data-transformation development process. Using DuckDB as the execution engine for running PySpark queries enables data practitioners to run large test suites over more data than would be otherwise possible within their local environment. This enables data teams to receive feedback on regressions earlier on in their development cycle as they can run unit tests, data validation tests, and integration tests that depend on PySpark query results with less friction in their local environment.

- Lastly, another effective quality assurance technique is to run your tests as part of a **continuous integration** (**CI**) process, where your data transformation code is required to be run automatically against the full test suite for any changes to be merged into the main code branch. An effective way to achieve this is to use a CI tool to spin up ephemeral environments where your test suite can be run. However, the inherent complexity of as distributed Spark cluster is that it's not well suited to being spun up rapidly as a short-lived resource. Alternatively, teams may test against long-lived clusters, but this introduces additional costs, and their state and resource contention would need to be managed. DuckDB, on the other hand, with both its small footprint and impressive performance, is well suited to running PySpark tests within CI environments, without you having to sacrifice test-execution time or the ability to run data validation tests over larger and more representative datasets.

At the time of writing, the DuckDB Spark API is made available as an experimental API that does not have complete feature coverage yet. We are excited about the additional use cases that the DuckDB Spark API will unlock; watch this space for updates!

In the next section, we'll focus on a specific type of use case for DuckDB. We'll look at different strategies for using DuckDB to drive user-facing data apps – applications that DuckDB is well suited to supporting.

DuckDB-powered data apps

There are many different types of data applications, but when people talk about *data apps*, they typically refer to purpose-built tools with interactive interfaces that enable users to support decision-making within a specific domain or enable self-serve data exploration and analysis. The growing popularity of data apps has been fueled by a proliferation of libraries and frameworks that can greatly streamline their development and enhance their capabilities, many of which remove the need for spinning up a whole frontend development stack.

Data apps include both data products and tools developed for internal stakeholders within an organization to support and enable their responsibilities, as well as those developed for end consumers who are external to an organization, such as customers, partners, or investors. When building data apps for end consumers, they often have different design needs, such as more polished interfaces and being able to scale to more users.

DuckDB is particularly well suited to powering data apps. First and foremost, DuckDB provides excellent support for the types of analytically oriented queries and data transformations that are frequently required by data apps, from complex statistical modeling, all the way to building and querying data cubes powering interactive BI dashboards. A key property of a successful interactive interface is responsiveness to user input. Therefore, the latency of query responses that are triggered by user input is an important consideration. DuckDB's blazing-fast speeds enable much lower latency of queries compared with other approaches that teams might reach for, such as using pandas dataframes or connecting to a PostgreSQL database. Lastly, DuckDB can consume a wide range of data sources and it also offers a great degree of flexibility in query patterns, supporting queries across multiple data types and locations simultaneously. This versatility enables DuckDB to support a wide range of data app use cases and a wide range of technology ecosystems they might need to be integrated into.

In this section, we'll look at two different strategies for using DuckDB to power your data apps. The first approach involves developing browser-based applications where DuckDB itself runs fully in the user's browser, integrating with web application code. This approach is well suited to building end-consumer-oriented applications as it exposes the full flexibility of the modern web platform, enabling developers with frontend development skillsets to build extremely customized solutions. The second approach involves using libraries that give data practitioners a unified framework for rapidly building client-server-oriented applications using Python or R, without exposing the full complexity of the modern web platform. This is particularly useful for enabling data teams to build prototypes and tooling for their workflows, as well as developing and publishing internal data apps for use across the organization by a range of stakeholders.

Building data apps with DuckDB-Wasm

WebAssembly (often shortened to **Wasm**) is a portable binary-instruction format for executable programs that are designed to enable high-performance web apps to be able to run in the browser. WebAssembly defines a low-level virtual machine that is supported by all modern browsers and runs in an environment that's isolated from the runtime of the host. Wasm modules can be imported by JavaScript applications, enabling them to interact with the API that's exposed by each module. Importantly, WebAssembly can be used as a compilation target for high-level programming languages. This means that a variety of languages, including C, C++, Rust, and even Python, can be compiled in a Wasm binary format to be run within web browsers at near-native speeds. This has enabled DuckDB (which is implemented in C++) to provide an official Wasm client, meaning that DuckDB can be run entirely within a browser, using the Wasm runtime.

The **DuckDB-Wasm** client (`https://duckdb.org/docs/api/wasm/overview`) enables the development of web applications that run DuckDB entirely on the client side in the browser – running at near-native speeds – bringing a modern high-performance analytical database to the browser. The integration of DuckDB with JavaScript and TypeScript application code bases enables the use of powerful frontend frameworks such as React and Angular to leverage both custom and pre-existing open source web components for interactive data visualization and analysis. This opens the possibility of creating rich, dynamic, and reactive user experiences for data apps, with DuckDB handling data ingest, data management, and query execution. Pushing compute to the client in this way – as opposed to running DuckDB server-side – can eliminate latency-inducing roundtrips to the server, dramatically improving the responsiveness of user interactions.

DuckDB-Wasm works by automatically offloading queries to dedicated worker threads and can read Parquet, CSV, and JSON files from the local filesystem, memory buffers, or from files hosted on HTTP servers and S3-compatible cloud object storage. Query results are returned as in-memory Apache Arrow tables, which means that no data copying is required for your web application code to consume them. When running queries, you have the option of fully materializing query results as an Arrow table, or you can lazily fetch result chunks through an Arrow stream reader. DuckDB-Wasm also supports prepared statements for parameterized queries.

The DuckDB-Wasm client provides several libraries that each offer different ways to integrate DuckDB into your web application. The duckdb-wasm package can be imported into your JavaScript or TypeScript code for deep integration with your application. For more advanced needs, you can also customize the package by building it from the source. DuckDB-Wasm also provides another library, duckdb-wasm-shell, which is an embeddable standalone DuckDB web shell. This provides a DuckDB CLI, very similar to the one we have been using in this book, that runs on Wasm, and which you can embed within your applications. Next, we're going to dive in and get hands-on with the DuckDB web shell in a web browser.

Exploring the DuckDB web shell

DuckDB provides a hosted version of the DuckDB web shell at https://shell.duckdb.org. Not only does it enable people to quickly take DuckDB for a spin without having to install anything, but exploring the hosted DuckDB web shell also helps convey a sense of the capabilities of DuckDB-Wasm. Let's go on a quick hands-on tour of the DuckDB web shell.

Start by navigating to https://shell.duckdb.org in your browser. You will be presented with an interactive DuckDB shell that closely resembles the DuckDB CLI shell, which the web shell has been designed to emulate:

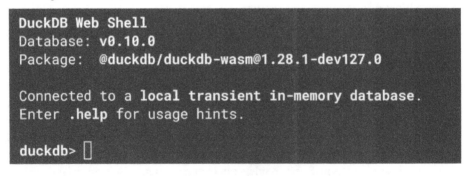

```
DuckDB Web Shell
Database: v0.10.0
Package:  @duckdb/duckdb-wasm@1.28.1-dev127.0

Connected to a local transient in-memory database.
Enter .help for usage hints.

duckdb> []
```

Figure 12.3 – The DuckDB web shell

The first thing we'll want to do is load some data for querying. DuckDB-Wasm provides a file web filesystem that can be used to query both local and remote files. However, as we are now operating from within a web browser, this functionality works a bit differently from how DuckDB normally reads from local and remote filesystems. When it comes to remote HTTP queries in the DuckDB web shell, it is possible to directly query remotely hosted files (as we saw in *Chapter 5*). However, an important caveat is that any remote web hosts you want to be able to query with DuckDB must be configured with **cross-origin resource sharing (CORS)** HTTP headers that enable requests from the domain that the DuckDB web shell is hosted on. This is a mechanism that's enforced by modern web browsers to prevent scripts from executing malicious code. See Mozilla's MDN documentation on CORS for more information: `https://developer.mozilla.org/docs/Web/HTTP/CORS`.

A reliable method to load data into the DuckDB web shell is to load files from your local machine into DuckDB-Wasm's web filesystem and then query them. This manual data-loading step involving user input is necessary due to the sandboxed execution model that modern web browsers are designed around, where applications are not given direct access to the underlying filesystem of the local machine, to prevent malicious content that might be encountered in the wild from being executed.

Before we load a file into the shell that we can then query, you might like to test that we are running DuckDB entirely within the browser. To do this, simply turn off your internet connection or put your device in airplane mode before continuing with this exercise. Since DuckDB is running entirely within the browser through the Wasm-powered DuckDB web shell, we won't be querying any remote files, meaning that everything we will do next can be performed offline.

Loading files into the DuckDB web shell

To load a file into DuckDB's web filesystem, we need to enter the following command:

```
.files add
```

After evaluating this command in the web shell's CLI, your web browser will open a file selection dialogue box. Navigate to the `chapter_12/countries.parquet` file located in the repository for this book and open it. Once added, this Parquet file will be registered in the virtual filesystem available to the DuckDB web shell. We can confirm this by issuing the following command, which will display a list of files currently registered:

```
.files
```

We'll get the following output:

```
duckdb> .files
```

File Name	File Size	Protocol	Statistics
countries.parquet	unknown	Http	false

```
duckdb> []
```

Figure 12.4 – The file we made available in the virtual filesystem of the browser

With the `countries.parquet` file now registered in DuckDB's web filesystem, we can now run DuckDB SQL queries, just as if we were using the DuckDB CLI. Let's query the `countries` file, limiting our results to the first 10 records returned.

```
FROM 'countries.parquet' LIMIT 10;
```

We'll get the following result:

```
duckdb> FROM 'countries.parquet' LIMIT 10;
```

country	name
AD	Andorra
AE	United Arab Emirates
AF	Afghanistan
AG	Antigua and Barbuda
AI	Anguilla
AL	Albania
AM	Armenia
AN	Netherlands Antilles
AO	Angola
AQ	Antarctica

```
duckdb> []
```

Figure 12.5 – The result of a query being run inside the DuckDB web shell

Once loaded into our Wasm-powered DuckDB session, we can continue to work with this data, uploading further files or directly querying files hosted on accessible web hosts (after you have restored network access), querying and transforming across data sources using any of DuckDB's core SQL operations – all being processed entirely within the browser.

When should you consider alternatives to DuckDB-Wasm?

For teams and practitioners building browser-based data apps, there are some considerations for when DuckDB-Wasm might not be the most appropriate choice. DuckDB-Wasm is designed with a batteries-included approach, aiming to give users the features of a full analytical database, embedded in the browser. A trade-off in achieving this is that the DuckDB-Wasm module and its required dependencies come with a reasonable footprint, resulting in increased latency from package fetching and loading. If very-low latency initial page load time is a key concern for your app's user experience, DuckDB in-the-browser may not be suitable for your needs. Similarly, if your data app is only handling a small number of records, you may not see any performance benefits from adopting DuckDB due to its initialization overheads.

Hopefully, this has helped illustrate how the DuckDB-Wasm client unlocks a powerful synergy: the versatility and analytical processing power of DuckDB, with the richness and ubiquity of the modern web platform. We are already seeing both open source and commercial services based around data apps that are built using DuckDB-Wasm. On the open source side, the Evidence project, which we covered earlier in this chapter in the *Tools and services for working with DuckDB* section, is using DuckDB-Wasm to enable the development of interactive dashboards and decision support tools that run entirely in the browser. This is an effective pattern that we believe we will be seeing increasingly more of for producing rich interactive analytical data apps.

Making data apps with Python and R

Another approach to building data apps powered by DuckDB is to develop server-side services that perform data processing in response to user interactions in a corresponding frontend interface. This approach offers more flexibility in how you choose to run and integrate DuckDB into your application on the server. This enables data practitioners to adopt their preferred language for working with data, which is frequently Python or R, along with all the data libraries from their toolkits, giving practitioners a more familiar development environment where they can be more effective. A challenge for data teams and practitioners is that building bespoke modern frontend applications involves specialized skillsets and can be resource-intensive to develop, requiring the establishment and maintenance of frontend-specific development toolchains and workflows that are very different from those found in most data teams. Unless there is a product need to develop quality end-consumer-facing data apps (in which case it may make sense to dedicate a frontend engineering team), this pathway is often not an appropriate option for many data teams as the resources that are required are too high for many of their target use cases, such as rapidly building agile prototypes and internal-facing tools and services.

There is a range of open source libraries that aim to address these challenges by enabling data practitioners and teams to develop interactive data interfaces in Python and R, without requiring frontend development skillsets. The libraries we'll look at here each take slightly different approaches from each other to enable data practitioners to build interactive data apps that are enabled through a web browser, but they do share some common high-level features. They all take the form of **reactive interfaces**, where user interactions trigger requests to a server process that performs necessary data processing for the query, returning results to the user interface, which is then dynamically updated accordingly. They enable users to configure both the data processing logic that happens on the server as well as the layout of the frontend user interface all within the same Python or R code, with the framework taking care of how to hydrate this code-as-configuration into a holistic data app composed of both a client and a server.

The abstractions that are provided by these libraries give data practitioners a streamlined experience for rapidly developing data apps, without the need for a supporting frontend development team, and without having to adopt a complex frontend development toolchain themselves. This enables domain-oriented ownership, where data apps can be developed by teams who have the most context around the domain and problems being solved, rather than development being dictated by the technical capabilities of teams. Here are some contexts where this is particularly beneficial:

- **Agile prototyping**: The ability to rapidly develop prototype interfaces and make changes without the need for hand-off to a web development team. This enables data teams to solicit feedback from a range of stakeholders, supporting iterative demonstration of value and course correction during the development of a project. The development of a tangible prototype is also a powerful vehicle for gaining buy-in from stakeholders and is often needed to secure funding for building out the app as a scalable and production-grade data product.

- **Internal data tooling**: Data projects frequently require a variety of bespoke supporting tools throughout their life cycle. This is particularly true of ML and data science projects, where custom tooling augmentation can be invaluable for data annotation, error analysis, and domain-specific data exploration. By giving data teams the ability to make ad hoc data apps, they are enabled to develop internal tools that meet the needs of their project.

- **Internal data products**: There are many opportunities for building data tools as published data products for wider use across an organization by a range of stakeholders, including customized reports, dashboards, and decision-support tools that support a range of business functions.

When building data apps for these types of use cases, they will frequently require an analytical data processing engine for running queries generated from user interactions. For example, when someone is using a dashboard, their selections of parameters, such as the date range of a chart, will result in a query being run against the relevant data source with appropriate filters applied. DuckDB is well suited for serving as the query engine that powers these kinds of apps for several compelling reasons:

- DuckDB's blazing fast performance on analytical workloads supports low-latency response times for user interaction, which is a crucial driver for a positive user experience.

- DuckDB's in-process execution model enables simple integration with data app code as all you need to do is import the DuckDB library. This eliminates the need to connect to and authenticate with a remote data source.

- DuckDB's versatility, from its comprehensive set of analytical functionalities to its ability to consume a wide range of data sources – all the features we have seen in our journey so far – means that it will be able to drive a range of different data app use cases, with diverse sets of needs.

Now, let's go on a quick tour of some notable open source libraries that are designed to help data practitioners rapidly develop data apps, and which DuckDB can be readily integrated with.

Jupyter Notebook workflows

When working within a Jupyter Notebook-based workflow, the **Ipywidgets** Python library (`https://ipywidgets.readthedocs.io`) and its extended ecosystem provides a rich range of interactive HTML widgets that can be used to add interactivity to data visualizations and workflows within Jupyter Notebooks. While this works well in the context of developing interfaces for use in analytical workflows, Jupyter Notebooks are not designed to be able to be deployed as shareable data apps. One issue, in particular, is that publishing a notebook as-is would involve giving users full access to the underlying kernel, effectively giving end users unrestricted access to the underlying machine, and therefore posing something of a security nightmare. The next three libraries we'll outline offer different approaches for safely building publishable interactive interfaces using Jupyter Notebooks, and all have good Ipywidget integration.

Voilà (`https://voila.readthedocs.io`) is a Python library that enables Jupyter Notebooks to be published safely as standalone data apps by exposing the desired interface alongside interactive components, without exposing the underlying Python kernel to users. Using Voilà to publish Jupyter Notebooks containing Ipywidget-based interactivity is an effective combination to rapidly build and share an interactive data app.

Panel (`https://panel.holoviz.org`) is another Python library that you can use to build interactive data apps using Jupyter Notebooks. Where Voilà focuses on providing a mechanism to convert notebooks into web apps, Panel provides a comprehensive dashboarding toolkit for building out customized user interfaces. Panel offers flexible development workflows for different contexts: you can use it to deploy individual notebooks as data apps (such as Voilà), or you can develop your data apps as standalone Python modules, without the need for Jupyter.

Solara (`https://solara.dev/`) is a pure-Python web framework that, like Panel, enables the development of data apps that can run both inside Jupyter Notebooks and as standalone Python modules. Similar to Panel, it also offers integration with Ipywidgets and its ecosystem. Solara is notable for being designed to support the development of larger and more complex apps. To achieve this goal, the framework has adopted a declarative interface that is based around reusable components that you can define yourself in Python or draw from the open source Ipywidgets ecosystem. Similar to JavaScript's React framework, which Solara draws inspiration from, Solara ensures that components take care of updating their own visual display based on the value of the data they currently have, which greatly simplifies state management, allowing you to focus more on the data flowing through your app rather than low-level details of how it should be displayed. Solara also offers a flexible range of deployment options for its apps; it can be run via a range of Python web framework libraries, including Flask, FastAPI, and Starlette, or it can even be incorporated into a Panel app, or deployed from a Jupyter Notebook using Voilà.

ML and AI data app development

As the development of ML and **artificial intelligence** (**AI**)-enabled data products has become an increasingly common activity of data teams across a wide range of organizations, the need for highly customized tools to augment use-case-specific ML workflows has also become increasingly apparent. This includes things such as data annotation tools to support supervised ML models, bespoke data exploration tools that can handle the high dimensionality associated with many ML and data science projects, and model monitoring diagnostic tools. Any of the data-app-building libraries we'll cover here can be used for these kinds of use cases, but the three libraries we'll mention now were specifically motivated with the tooling needs of ML and data science teams in mind.

Streamlit (`https://streamlit.io`) is a Python library for rapidly building sharable reactive web apps out of Python scripts and, at the time of writing, is likely the most well-known Python library for building data apps. It was designed specifically with the needs of data scientists and ML engineers in mind, enabling them to build bespoke data apps that support a range of activities across the life cycle of ML and data science projects, from data collection all the way to prototype development. More recently, the Streamlit project has broadened its focus around building sharable data apps for a range of contexts beyond ML and data science. If you want to learn more about developing data apps with Streamlit, Packt's book *Streamlit for Data Science: Second Edition*, by Tyler Richards, is a great resource.

Gradio (`https://www.gradio.app`) is another Python library that has been specifically designed to allow data apps to be built for ML and AI projects. Gradio emphasizes supporting the development of sharable data apps that can be used as interactive prototypes of ML models or data science workflows. It also aims to support the development of tools that support the development of these data products through bespoke interfaces for debugging, testing, and analysis. It has become a popular choice in the ML community for sharing interactive demos of a wide range of different types of models and data resources, notably within the Hugging Face community (`https://huggingface.co`).

Wave (`https://wave.h2o.ai`) is a library for rapidly developing ML and data-analytics-oriented apps, again with a particular emphasis on performance and low-latency responses, to enable real-time analytical use cases. It currently has bindings for building Wave apps in both Python and R. Its underlying content server, however, is written in Go, and compiles without any dependencies. As a result, it can be deployed on a wide range of operating systems and offers extremely fast performance, making it well suited to developing real-time analytics dashboards, in which data visualizations are updated live from streaming data. A notable feature of Wave is that it provides integrations for single-sign-on support through an organization's authentication provider. There is also a Visual Studio Code extension offering for further acceleration of app development through a contextual library of Wave code snippets, including app skeletons and layout recipes, such as headers and sidebars.

General-purpose analytical data apps

Shiny (`https://shiny.posit.co`) is the go-to library for building data apps in R, and more recently. it has also gained support for Python. It is also, in many ways, the original data app-building tool for data scientists and data analysts, with many of the other tools mentioned here drawing inspiration from Shiny. It's a great general-purpose tool for building different types of data apps, including sharable prototypes, published dashboards and reports, and ad hoc data tool development.

Dash (`https://dash.plotly.com`) is a Python library for rapidly building data apps in Python. Its backend server is built on Flask, a popular Python web framework, making it readily deployable across a range of environments. In addition to making standalone web apps, Dash can also be used within Jupyter Notebooks. Dash comes with a built-in set of core components for building interactive data apps, including strong integration with the Plotly visualization library, which we used in *Chapter 11* to support exploratory data analysis. Dash is also notable for having a high degree of customizability in terms of appearance, with its component-based layout specification offering first-class support for **Cascading Style Sheets** (**CSS**). While developers of Dash apps can work entirely in Python, since Dash is based on top of the frontend library React, it can be readily extended by converting existing open source React components into Dash components. This has enabled a rich ecosystem of community-supported Dash components that support a wide variety of functionality. Dash's design principle of stateless app execution also means that, if needed, the backend service behind Dash apps can be horizontally scaled across multiple worker processes or compute instances. This, combined with Dash's strong support for visual customization, makes Dash suitable for developing more scalable data products that can be deployed to a wider audience across an organization, such as BI dashboards and reports.

Selecting a DuckDB-powered data app strategy

In this section, we introduced the idea of using DuckDB to power the analytical querying capabilities of a range of different types of data apps. We outlined how DuckDB's powerful analytical capabilities, combined with its performance and simplicity of integration, make it a natural fit as a query engine that's used to drive compelling interactive analytical interfaces. Hopefully, you too are excited by the possibilities of DuckDB-powered data apps, though you might still be wondering where to start, given the wide range of approaches and open source tools covered in our whirlwind tour. We will conclude this section by summarizing the key information we presented, along with some heuristics to help you identify candidate strategies more likely to be appropriate for a given use case.

The two broad approaches for building data apps with DuckDB that we covered are as follows:

- Building rich data apps that run entirely within the browser using the DuckDB-Wasm client

- Making data apps in Python and R using one of the many open source libraries that allow data practitioners to define both user interface components and user interaction handling logic within the same code

When choosing which of these approaches is most appropriate for a candidate data app, it's important to consider both the needs of consumers of your app, as well as the developers and ongoing maintainers of the app, as different use cases may need to prioritize needs of these respective personas differently.

Data apps that are closer to end-user-facing products will tend to call for higher degrees of responsiveness, availability, and scalability to a larger audience, as well as general polish and close attention to user experience. For such contexts, you may want to consider developing a web app that uses DuckDB-Wasm to run queries triggered by user interactions directly in the browser. While this approach does tend to require more investment in terms of both team capability and capacity, harnessing the benefits of client-side execution and the power of the modern web platform may be necessary to enable your target user experience. Concretely, the benefits of building in-browser apps using DuckDB-Wasm are as follows:

- **Lower-latency response times**: Running DuckDB in the client eliminates costly roundtrips to a remote server for query execution.

- **Out-of-the-box availability**: The distribution channel for your data app is essentially just a web browser, which your users will already have access to across a range of devices, including desktops, tablets, and phones.

- **Scalability**: The client-side execution model means that you can make much more effective use of the computational resources available on your local devices. Pushing query execution to the client frees you from having to scale these operations horizontally within your infrastructure by reducing or even eliminating the need for remote query execution.

- **Customizable and rich user experiences**: The modern web platform offers a powerful and flexible base to build the kinds of customized product-specific features and rich user experiences required to realize the value of your application. Notably, the teams that are developing your data app will also be able to leverage a wide range of open source libraries as accelerators, including web frameworks for building the app with, as well as libraries that provide rich web components, such as charting libraries for rich data visualizations, and prebuilt controls for users to interact with your app.

On the other hand, there is a range of contexts where it may be more important to prioritize other needs around the app development side, such as the skills that will be required to build and maintain it, the development time of new features, and the domain knowledge required to maintain it. This is particularly worth considering for data apps that have been designed for use internally. Here are some examples:

- Internal data products such as reports, dashboards, and custom decision-support tools

- Interactive demos and proof-of-concept apps for data science and data analytics initiatives that enable agile prototyping

- Custom data tooling to support frequent workflows found in data science, ML, and data analytics teams

For such contexts, it may be worth considering a strategy that enables your data teams to develop and maintain data apps themselves by adopting a tool that allows data teams to use tooling that they are familiar with and can operate themselves. This is where combining DuckDB with the kinds of data-app-building tools for Python and R that we have discussed is worth considering. More concretely, here are some factors that may support this strategy:

- **Skillsets and personas**: Many data teams are unlikely to have the frontend engineering skills and experience required to build and maintain a modern web application. For data teams that do have these capabilities, the overheads involved in spinning up and maintaining frontend projects and their supporting tooling will often make this option non-viable for them to take on themselves, given their existing responsibilities. Choosing to build an internal data app using DuckDB-Wasm will often require involving an engineering team with the requisite capability and capacity – which, in turn, requires that funding for this team is both available and considered a worthy investment.

- **Cycle time of development**: Common to all the internal use cases we have called out is that their delivery is frequently time-sensitive. Decision support tools need to be updated to reflect dynamic changes in business needs and environmental circumstances, effective development of proof-of-concept apps requires that changes can be made rapidly in response to early feedback, and when a data team needs new tooling functionality for data preparation or analysis, their work may be blocked while it is being developed. If a data team requires the involvement of an external team for the development or enhancement of a data app that they are responsible for, the total time taken to deliver new features from start to end (the cycle time) will be significantly increased by the introduction of a handover point in the work, which inevitably leads to time waiting.

- **Domain-oriented ownership**: Data apps that function as internal decision-support tools tend to coincide with a high degree of specialized domain knowledge. Consider a dashboard for sales and financial reporting or a custom interface that exposes a customer-churn forecasting model. Adopting a technology-oriented ownership model, where these tools are maintained by engineering teams by virtue of their skillsets, takes ownership of the data product away from the people who have the most context around the requirements and needs of the app, and who are best placed to ensure that new features will be fit-for-purpose. For data apps that play key business-supporting roles, it is frequently desirable to enable as much domain-oriented ownership as possible.

For many data teams and individual data practitioners, these considerations will mean that owning, developing, and maintaining data apps themselves, rather than involving an external engineering team, will be the preferred option. This is where using a Python or R library that's designed to support data app development, in conjunction with DuckDB, can be a powerful enabler. If you've decided that this is the right pathway for you, but are unsure which might be the right data app tool to adopt, here are some heuristics you can use to get started:

- If your team primarily uses R, it's hard to go wrong with **Shiny**, the go-to tool for building data apps with R. Of the tools we called out in this section, Wave is the other library with R support and could be worth considering for data apps with strong real-time analytics needs. The following suggestions are all Python-oriented.

- For teams who make heavy use of Jupyter Notebooks for their data analysis workflows, the Ipywidgets library and its rich ecosystem are worth exploring. The Ipywidgets library itself, along with an extensive range of open source widget libraries, can enhance Jupyter Notebook-based development loops with visualization components and user-input controls that can be used to streamline customized analytical workflows. We then saw several libraries geared toward developing Ipywidget-based interfaces into data apps that can be shared and published. The Voilà library is the simplest of these, allowing you to easily convert Jupyter Notebooks with interactive components into deployable web apps. For more customized dashboard-like interfaces, and for escaping the confines of the notebook, Panel and Solara are two Ipywidget-compatible data app frameworks. Panel's simple approach to defining interface layouts may be more suitable for practitioners with less experience working with web technologies, whereas Solara, with its expressive React-inspired API, may appeal to practitioners with web development experience or when more customizability is required.

- For data teams who need to get started building their own customized tooling interfaces, any of the tools can be used effectively for this purpose. However, Streamlit stands out as being specifically designed for this use case, with its streamlined scripting API allowing practitioners to quickly spin up bespoke data apps with clean interfaces. Additionally, leveraging interactive Ipywidgets is also a compelling approach for this use case.

- If you are targeting a data app that requires considerable customizability, has more inherent complexity, or needs to be scaled to a larger audience, Dash and Solara stand out as being worth exploring. Both these libraries provide data practitioners with an abstraction that allows you to compose data apps as layouts of reusable components, and which can be freely styled using custom CSS. Both libraries also allow you to define custom components and leverage open source components from other ecosystems. Both Dash and Solara offer considerable flexibility around deployment options, with Dash being based on the Flask Python web framework and Solara allowing you to run on several Python web frameworks, such as Flask, Starlette, and FastAPI. The stateless design of Dash and Solara means that they can both be scaled horizontally across multiple worker engines, allowing your data apps to be served to a much larger number of users.

- If your goal is to build a prototype for an ML-powered data product, any of the tools we have discussed will help you rapidly bootstrap and iterate on an agile prototype. However, Gradio, Wave, and Streamlit are all tools that have been designed with ML applications in mind, so they may be worth assessing to see if any of their features will give you an enhanced accelerator for your use case.

These suggestions are simply a starting point; you may need to experiment with a few libraries to determine which ones meet your needs. It is also worth noting that the collection of open source tools for rapidly building data apps we have presented in this section is not exhaustive, and the number of tools in this space is growing continually.

We hope this has gotten you excited about the wide-ranging possibilities that DuckDB-powered data apps enable. In the next section, we'll conclude our DuckDB adventures by covering the various ways you can keep up to date with DuckDB developments and engage with the wider DuckDB community.

Keeping up to speed with the DuckDB ecosystem

DuckDB is attracting much attention and excitement within the data community, for all the reasons we have seen throughout our adventures in this book. It is also a relatively new project in the scheme of things, and both the DuckDB project itself and the ecosystem that it sits in are moving very fast.

While we are writing this (May 2024), the latest DuckDB version is 0.10.2. DuckDB 0.10 *Fusca*, released in February 2024, was the first DuckDB release not marked as having a preview status, indicating the maintainers' confidence in the maturity that DuckDB has reached. This is also evidenced by the number of organizations already using DuckDB operationally across a range of applications. The DuckDB maintainers have indicated that they are focusing on stability and robustness enhancements ahead of the upcoming 1.0 release. The notable feature that's anticipated to land in 1.0 is a stabilized storage format, which will enable future DuckDB versions to read and write databases produced by previous DuckDB versions. Besides this noteworthy milestone, the project is seeing fast-paced development around core DuckDB capabilities and performance, its extensions, and enhancements across its extensive suite of client APIs. Meanwhile, tooling in the data ecosystem is advancing. We have covered a decent slice of the currently available integrations with DuckDB in this chapter, but it is by no means exhaustive, and the ecosystem will certainly continue to grow.

You may be wondering how you can keep up to speed with the fast-moving developments in the DuckDB project and its ecosystem. As we come to the end of our DuckDB adventures, we will leave you with some breadcrumbs that you can follow to carry on discovering and learning new techniques and tools for working with DuckDB.

duckdb.org

The home of the DuckDB project is located at `https://duckdb.org`. This contains DuckDB's documentation and covers both core features and all its client APIs. This should be your first port of call for answering questions about DuckDB's capabilities and how to go about using them. You can also access the DuckDB's blog here, which is a valuable resource, with articles covering in-depth treatments of new DuckDB features and discussions of design principles that the project has adopted. We very much recommend reading through the archives if you have yet to explore them.

DuckDB's home page also provides information about upcoming and previous **DuckCon** events, which are in-person events that are open to the wider DuckDB community. Attendees of DuckCon hear about updates from the DuckDB development team, as well as talks from the community, sharing how people and organizations are adopting DuckDB to solve their data needs. Recordings of DuckCon presentations are made available by the official DuckDB YouTube channel. Additionally, you will find information about DuckDB's governance structure, including the DuckDB Foundation, and ways for organizations to become involved and support DuckDB development.

GitHub

The DuckDB project is hosted on GitHub (`https://github.com/duckdb/duckdb`). Here, you can follow active development of DuckDB features, submit bug reports when you encounter issues, contribute enhancements via pull requests, and discuss problems or puzzles you're having using DuckDB, as well as suggest potential DuckDB feature enhancements.

The DuckDB Discord server

The DuckDB Discord server (`https://discord.duckdb.org`) is a place where DuckDB users and developers come together to discuss all things DuckDB, including technical questions on how to do things with DuckDB, potential DuckDB enhancements, and share DuckDB-related projects and creations. It's a particularly valuable resource that's open to anyone with an interest in DuckDB. Do remember that community members are volunteering their time to help with questions, so before posting, make sure you've checked the relevant part of the DuckDB documentation to see if the information you need is there. The DuckDB documentation is being updated and expanded continually, so it can be worth checking back in now and then.

Awesome DuckDB

Awesome DuckDB (`https://github.com/davidgasquez/awesome-duckdb`) is a GitHub repository containing a curated list of DuckDB resources, maintained by David Gasquez. It contains links to a wide range of DuckDB resources, including many of the open source libraries we have covered in this chapter, and much more, across the following areas:

- DuckDB resources
- Tools powered by DuckDB
- Web clients
- Libraries powered by DuckDB
- SQL clients and IDEs that support DuckDB
- Projects powered by DuckDB
- Integrations
- Media – talks, podcasts, and blog posts

Awesome DuckDB is an excellent collection of resources for discovering further libraries and projects involving DuckDB, as well as informational content about DuckDB and the kinds of applications people are using it for. If you have a suggestion for a DuckDB resource that's not included in the list, contributions are welcome.

Stack Overflow

A resource that will be familiar to many already is Stack Overflow (`https://stackoverflow.com`), a popular online community where developers and technologists ask and answer questions. There is a growing number of DuckDB questions and answers, all of which can be found via the `duckdb` tag.

Social media

Active DuckDB discussions and information sharing can also be found on social media platforms, in particular LinkedIn and X (formerly Twitter). On both platforms, people use the #duckdb hashtag, which you can use to find people posting about DuckDB content, such as blog posts, conferences, webinars, and both local and online meetup events. If you look for and follow the official DuckDB accounts on LinkedIn and X, and start following other people posting about DuckDB, you should start seeing interesting DuckDB content popping up in your feeds.

Chapter and book summary

In this chapter, we covered a range of tools and services that can be used to augment and uplift DuckDB-oriented workflows – both for data analysis and software development – as well as online resources you can use to continue exploring and learning about DuckDB.

We hope you enjoyed this last chapter of *Getting Started with DuckDB* and our coverage of DuckDB, a modern in-process database for working with analytical data. We also hope that throughout this journey, as we have covered a wide range of DuckDB's features and capabilities, we have been able to share our enthusiasm for the versatility and power of this wonderful analytical data tool, as well as give you a sense of different applications and contexts it shines in: from using DuckDB to supercharge your data science and data analytics workflows to being a core component in interactive data products and lean data engineering infrastructure.

We started our journey in *Chapter 1* by unpacking what type of data tool DuckDB is, situating it in the database landscape, and identifying the types of analytical use cases it is well suited. We also set up the DuckDB CLI client for use in our hands-on exploration in the following chapters, as well as spending some time on a brief SQL primer for those who could benefit from a refresher.

In *Chapter 2*, we began our exploration of DuckDB's features, looking at how to load data into DuckDB from a range of formats and shapes, across CSV, JSON, and Parquet files. We saw how DuckDB's CSV reader and powerful CSV sniffer can automatically infer the structure of many CSV files found in the wild. We looked at a practical example of ingesting into DuckDB a public dataset from a bike share program, which is made available in CSV format, before ending the chapter with a brief look at exporting data from DuckDB into different file formats.

In *Chapter 3*, we explored DuckDB's powerful data wrangling capabilities, focusing in particular on how we can use SQL operations and language features to perform common transformations used in data analysis. We did this through two applied use cases: analyzing user behavior from web server logs and analyzing taxi trip data from New York City. We covered a range of operations and language features of SQL, including regular expressions, aggregation operations, window functions, and common table expressions.

In *Chapter 4*, we learned about DuckDB operations and performance, as well as how to work with timestamp data. We unpacked DuckDB's built-in indexes and how we can manually create indexes on columns to improve query performance. We also covered how to leverage DuckDB performance optimizations to improve file-reading performance through the use of Parquet files and Hive-partitioned data. We ended this chapter with a brief coverage of how DuckDB represents timestamp data and how to work with it effectively by using time series data representing the events of the Apollo 11 mission.

In *Chapter 5*, we introduced DuckDB's extension system, which enables users to extend DuckDB's functionality with useful features and capabilities that sit outside of the core DuckDB API. We covered how to install and load extensions before getting hands-on with several extensions, including the SQLite extension, the Full-Text search extension, and the Spatial extension. We concluded this chapter by getting hands-on with the Spatial extension, using it to perform operations on geospatial data to explore wine regions of France.

Chapter 6 saw us looking at a selection of DuckDB's features for modeling, generating, manipulating, and querying semi-structured data. We explored DuckDB's nested data types and functions for working with them. We also looked at DuckDB's strong support for working with JSON data, made available through its JSON extension, before finishing with some examples of using DuckDB to retrieve and analyze JSON data from web APIs.

The second half of this book saw us move away from using the DuckDB CLI and start working with the DuckDB Python client. In *Chapter 7*, we set up our Jupyter-Notebook-based IDE for working with DuckDB in Python and then unpacked the different ways to connect to DuckDB databases: using the convenient default in-memory database and the flexibility of explicitly creating connections, allowing us to configure and connect to both in-memory and persistent databases.

Then, in *Chapter 8*, we dived deeper into working with DuckDB's Python API, seeing how the client is well suited for both analytical workflows and developing software packages that leverage DuckDB. We covered the Python client's Relational API and its DB-API, as well as exploring a range of compelling integrations with the Python language and Python data packages, including pandas and Polars.

In Chapter 9, we turned to exploring DuckDB's R client. We covered a range of different ways you can leverage DuckDB in your R-based analytical workflows, including the lower-level DBI interface for interacting with DuckDB databases, querying and loading R dataframes into DuckDB, and converting queries and tables into dataframes. We also saw we can use the elegant and powerful dplyr interface to define DuckDB queries without needing to write SQL.

In *Chapter 10*, we surveyed some of DuckDB's SQL enhancements and features that are designed to improve the experience of writing analytical queries. This included looking at some of DuckDB's friendly SQL syntax, including affordances for easier column selection, inserting columns into tables by name, and function chaining. We also saw how DuckDB's positional join allows us to leverage the ordering of rows when joining tables, as well as how DuckDB's ASOF join enables the effective joining of temporal data.

In the last of our hands-on chapters, *Chapter 11*, we dove deeper into the practical side of things, looking at how we can use DuckDB to perform exploratory data analysis with Python. We started by looking at some tools that complement DuckDB well when performing data analysis with Jupyter Notebooks. We proceeded to explore the Melbourne Pedestrian Counting System dataset, using DuckDB to uncover some interesting observations about the historical movements of pedestrians throughout the central Melbourne region.

Finally, in this chapter, we concluded our journey into DuckDB by exploring aspects of the wider DuckDB ecosystem. This included tools and services that can assist you with analytical workflows involving DuckDB, such as SQL IDEs, tools for rapid data exploration, and alternative query interfaces that target DuckDB, which can aid with the portability of query code. We also looked at how DuckDB is well suited to powering analytical data interfaces and how there are a range of pathways for developing data apps that use DuckDB. Lastly, we provided some breadcrumbs for you to follow that will help keep you up to speed with the DuckDB project, discover further DuckDB resources and projects, as well as get you started participating in and contributing to the wonderful community that has formed around DuckDB.

This concludes *Getting Started with DuckDB*. We hope that it has sparked some of the same excitement we have for DuckDB within you and that you have been able to use it as a springboard to jump off and continue your adventures working and playing with DuckDB. Thanks for coming on this journey with us!

Index

A

adaptive radix tree (ART) **82**
aggregate functions **68, 69**
analytic functions **78**
Apache Arrow **320**
Apache Arrow table
  converting to **209**
  registering, as virtual table **236-238**
Apache Superset
  URL **316**
application programming
    interfaces (APIs) **125**
  JSON data, querying from **152, 153**
Arrow Database Connectivity (ADBC) **320**
ASOF
  temporal joins, using with **267-271**
atomicity, consistency, isolation,
    and durability (ACID) **5**
automated field updating **264**
Awesome DuckDB
  reference link **345**
AWS S3 **112**

B

Binder
  URL **318**
binning **98**
BI tools **315, 316**
block range index (BRIN) **82**
box plot **305**
  reference link **305**
business intelligence (BI) **7**

C

Cascading Style Sheets (CSS) **339**
change data capture (CDC) tools **317**
Cloud Storage **112**
columns
  selecting, effectively **256-260**
command-line interface (CLI) **12, 312**
comma-separated values (CSV) **23**
common table expression (CTE) **70, 116, 269**
Comprehensive R Archive
    Network (CRAN) **324**
continuous integration (CI) **330**
Coordinated Universal Time (UTC) **61**

packtpub.com

Subscribe to our online digital library for full access to over 7,000 books and videos, as well as industry leading tools to help you plan your personal development and advance your career. For more information, please visit our website.

Why subscribe?

- Spend less time learning and more time coding with practical eBooks and Videos from over 4,000 industry professionals

- Improve your learning with Skill Plans built especially for you

- Get a free eBook or video every month

- Fully searchable for easy access to vital information

- Copy and paste, print, and bookmark content

Did you know that Packt offers eBook versions of every book published, with PDF and ePub files available? You can upgrade to the eBook version at packtpub.com and as a print book customer, you are entitled to a discount on the eBook copy. Get in touch with us at customercare@packtpub.com for more details.

At www.packtpub.com, you can also read a collection of free technical articles, sign up for a range of free newsletters, and receive exclusive discounts and offers on Packt books and eBooks.

Other Books You May Enjoy

If you enjoyed this book, you may be interested in these other books by Packt:

Vector Search for Practitioners with Elastic

Bahaaldine Azarmi, Jeff Vestal

ISBN: 978-1-80512-102-2

- Optimize performance by harnessing the capabilities of vector search
- Explore image vector search and its applications
- Detect and mask personally identifiable information
- Implement log prediction for next-generation observability
- Use vector-based bot detection for cybersecurity
- Visualize the vector space and explore Search.Next with Elastic
- Implement a RAG-enhanced application using Streamlit

Mastering PostgreSQL 15

Hans-Jürgen Schönig

ISBN: 978-1-80324-834-9

- Make use of the indexing features in PostgreSQL and fine-tune the performance of your queries
- Work with stored procedures and manage backup and recovery
- Get the hang of replication and failover techniques
- Improve the security of your database server and handle encryption effectively
- Troubleshoot your PostgreSQL instance for solutions to common and not-so-common problems
- Perform database migration from Oracle to PostgreSQL with ease

Packt is searching for authors like you

If you're interested in becoming an author for Packt, please visit authors.packtpub.com and apply today. We have worked with thousands of developers and tech professionals, just like you, to help them share their insight with the global tech community. You can make a general application, apply for a specific hot topic that we are recruiting an author for, or submit your own idea.

Share Your Thoughts

Now you've finished *Getting Started with DuckDB*, we'd love to hear your thoughts! Scan the QR code below to go straight to the Amazon review page for this book and share your feedback or leave a review on the site that you purchased it from.

https://packt.link/r/1-803-24100-4

Your review is important to us and the tech community and will help us make sure we're delivering excellent quality content.

Download a free PDF copy of this book

Thanks for purchasing this book!

Do you like to read on the go but are unable to carry your print books everywhere?

Is your eBook purchase not compatible with the device of your choice?

Don't worry, now with every Packt book you get a DRM-free PDF version of that book at no cost.

Read anywhere, any place, on any device. Search, copy, and paste code from your favorite technical books directly into your application.

The perks don't stop there, you can get exclusive access to discounts, newsletters, and great free content in your inbox daily

Follow these simple steps to get the benefits:

1. Scan the QR code or visit the link below

https://packt.link/free-ebook/978-1-80324-100-5

2. Submit your proof of purchase
3. That's it! We'll send your free PDF and other benefits to your email directly

www.ingramcontent.com/pod-product-compliance
Lightning Source LLC
Chambersburg PA
CBHW080611060326
40690CB00021B/4657